BIRDS of the
WEST INDIES

G. Michael Flieg and Allan Sander

B L O O M S B U R Y
LONDON • OXFORD • NEW YORK • NEW DELHI • SYDNEY

POCKET PHOTO GUIDE

Bloomsbury Natural History
An imprint of Bloomsbury Publishing Plc

50 Bedford Square   1385 Broadway
London   New York
WC1B 3DP   NY 10018
UK   USA

www.bloomsbury.com

BLOOMSBURY and the Diana logo are trademarks of
Bloomsbury Publishing Plc

First published by New Holland UK Ltd, 2000 as *A Photographic Guide to Birds of the West Indies*
This edition first published by Bloomsbury, 2017

British Library Cataloguing-in-Publication Data
A catalogue record for this book is available from the British Library.

Library of Congress Cataloguing-in-Publication data has been applied for.

ISBN: PB: 978-1-4729-3814-5
ePDF: 978-1-4729-3812-1
ePub: 978-1-4729-3815-2

2 4 6 8 10 9 7 5 3 1

Designed and typeset in UK by Susan McIntyre
Printed in China

MIX
Paper from
responsible sources
FSC® C008047
www.fsc.org

To find out more about our authors and books visit www.bloomsbury.com.
Here you will find extracts, author interviews, details of forthcoming events
and the option to sign up for our newsletters.

CONTENTS

INTRODUCTION

Birds of the West Indies illustrates over 250 of the region's more common species which either occur nowhere else in the world (endemic species) or are present in the region throughout the year or for a large part of it. Those species which simply migrate through the West Indies are not included, nor are rarer mainland species which wander here infrequently. Instead, the book concentrates on those species which are likely to be encountered by birdwatchers. The small size of this book enables one to carry it in a pocket for instant retrieval. Almost all of the photos have been taken in the field in the birds' natural habitats.

There have been over 560 species recorded in the West Indies. Some species have become extinct (Grand Cayman Thrush), some have been rediscovered (Puerto Rican Nightjar), and others are known from but a few occurrences.

The West Indies are ideal for birdwatching. While exploring in a port of call, or simply vacationing in the sun, there is always some time available to explore and observe an island's birds. Each island presents a unique association of characteristic birds, many of which are endemic. If you have some free time, there may be adequate opportunity for a thorough introduction to the birdlife of a single island or group of islands.

The 250 species illustrated here are the more common and conspicuous birds of the West Indies; nearly all of the families found in this area are included. Also shown are some of the more rare, interesting and spectacular species. By studying these pictures carefully you should be able to identify a species when you encounter it. This book is not a complete guide, and pictures are not always sufficient for identification in the field. Behaviour, songs and views from different angles are all important, and a single photo is sometimes inadequate for proper identification. Notes and sketches can be important when using another sourcebook, at a later time, to identify a species. Often a single characteristic is enough to be certain of a bird's identity, but it usually requires a combination of features to distinguish one species from another one which is closely allied to it.

Some birdwatching can be done without equipment. Many birds come to feeders, picnic tables and lawns, and are relatively tame. A pair of binoculars is an important tool for any birder, and their costs vary widely. Research the products on the market carefully for the one that fits your needs and budget.

Birds of the West Indies, by James Bond, was published in 1936. This book underwent many printings and several slight revisions, but in the end was unable to keep pace with advances in the ornithology of this region; by 1970 it had become outdated. It was not until 1975 that another field guide appeared, but it was specific to the Bahamas. In 1985 a Cayman Islands guide was produced, and a Puerto Rican and Virgin Islands guide followed in 1989. *Birds of the Eastern Caribbean* and *Birds of Jamaica* were published the following year. But still there was no guide to all the birdlife of the West Indies to replace Bond's. This void was filled in 1998 with the publication of *A Guide to the Birds of the West Indies*, compiled by the most prominent ornithologists of their respective areas. This book is comprehensive, and has raised the West Indian field guide to a new and higher level. Its publication has rendered Bond's volume obsolete, but we should never underestimate the importance of a work which stood the test of time for over 60 years. The new guide has incorporated species recently discovered or split from others; distribution ranges have been newly defined; it is well illustrated with paintings, and stands on its own.

HOW TO USE THIS BOOK

This book has been designed for clarity and with ease of use in mind. On page 6 is a key to symbols used on each page of the main species descriptions. These symbols will guide the reader to a family or group of families to which each bird belongs. Each such symbol appears on the first full page bearing descriptions of that group. Photographs show the commonly seen plumage. Where there are two or more photographs they are generally of male and female, or adult and immature. The male or adult is usually on the top or left. Other captions are explained in the text.

The species descriptions follow the American Ornithologists' Union's *Check List of North American Birds*, 7th edition (1983) for the taxonomic order, as well as for the scientific and common names of West Indian species. There have been a few changes to names based on recent researches. Each species description begins with the common name, scientific name and length of the living bird from bill tip to tip of tail. The next few sentences describe the main identification features. This is followed by a paragraph detailing range, habitat, voice, food, behaviour, and other features pertaining to the species. A glossary of terms used in the descriptions is given on page 139, and the diagram below shows the names applied to the different parts of a bird.

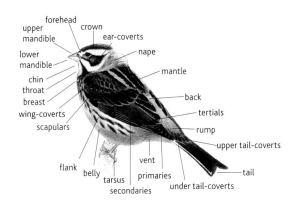

forehead
upper mandible
crown
ear-coverts
lower mandible
nape
chin
throat
breast
mantle
wing-coverts
back
scapulars
tertials
rump
upper tail-coverts
flank
belly
tarsus
vent
secondaries
primaries
under tail-coverts
tail

KEY TO COLOURED TABS

Grebes to frigatebirds

Herons, storks & vultures

Ducks

Raptors

Rails, crane & limpkin

Waders

Gulls & terns

Pigeons & doves

Parrots & parakeets

Cuckoos & ani

Nightbirds

Hummingbirds

Trogons

Todies & kingfishers

Woodpeckers

Flycatchers

Vireos

Crows & swallows

Wrens & gnatcatchers

Thrushes

Shrikes & shortwings

Wood warblers

Tanagers, euphonias & bananaquit

Bullfinches & grassquits

Sparrows

Blackbirds & orioles

Crossbill & siskin

THE WEST INDIES AND ITS ENDEMIC BIRDS

The West Indies includes the Bahamas, Greater Antilles (comprising the islands of Puerto Rico, Jamaica, Hispaniola and Cuba), Lesser Antilles, Cayman Islands, Virgin Islands, San Andreas and Providencia. The islands of Bermuda, the Netherlands Antilles, Trinidad, Tobago and Cozumel are not included.

With the exception of the Bahamas and Barbados the West Indies are of volcanic origin, and erupted from the ocean floor; this has allowed the islands to develop a unique birdlife over time, and it is one of the greatest centres of endemism in the world. Many species are unique (endemic) solely to one or several islands. The birds of the West Indies do not appear to be mainly of South American origin, as are those of Trinidad and Tobago, but instead many endemic species have evolved due to isolation from the mainland.

There are about 110 single-island endemics, and an additional 70 or so multi-island endemics. There are only two endemic bird families in the West Indies: the todies (Todidae) of the Greater Antilles and the Palm Chat (in a family of its own, Dulidae) on Hispaniola.

Most bird species of the West Indies are migratory, arriving on the islands from North America, and then either remaining or continuing their journey southward. The return migration is usually the reverse of this. Some species overfly the Atlantic from north-east North America directly to the West Indies and northern South America, bypassing landfall in the United States.

The IUCN (International Union for Conservation of Nature and Natural Resources) has established a classification for birds which are threatened on a global scale, and the criteria for this have been applied in individual cases by BirdLife International. In the lists below, such globally threatened birds are indicated as follows:

CE Critically Endangered: facing extinction in the wild in the immediate future.
E Endangered: not critical but facing a high risk of extinction in the near future.
VU Vulnerable: not critical or endangered, but faces a high risk of extinction in the medium term.
NT Near threatened.

Bahamas – 3 single-island endemics

Bahama Woodstar
Bahama Swallow
Bahama Yellowthroat

The swallow breeds in the Bahamas, but many winter in north-east Cuba. The Giant Kingbird which formerly occurred in the Bahamas is now extinct there and is confined to a small area of central Cuba. An accidental introduction of Cuban Grassquit into New Providence in 1963 led to it becoming established there. The Bahama Woodstar is rare in Grand Bahama due to competition with Cuban Emerald.

Andros, Grand Bahama and Nassau are the largest of the Bahamas, a group of islands lying off the south-east coast of the United States. Abaco has a population of Rose-throated Parrots which have learned to breed underground. A huge colony of Greater Flamingos breeds on Great Inagua. Other islands of the group include Bimini, Exuma, Eleuthera, Cat Island, San Salvador, Long Island, Crooked Islands, Mayaguana and many other small scattered cays. The Turks and Caicos are included as well.

Puerto Rico – 16 single-island endemics

Puerto Rican Parrot (CE)
Puerto Rican Lizard-cuckoo
Puerto Rican Screech Owl
Puerto Rican Nightjar (CE)
Green Mango
Puerto Rican Emerald
Puerto Rican Tody
Puerto Rican Woodpecker
Puerto Rican Pewee
Puerto Rican Flycatcher
Puerto Rican Vireo
Elfin Woods Warbler (VU)
Puerto Rican Stripe-headed
    Tanager
Puerto Rican Tanager
Puerto Rican Bullfinch
Yellow-shouldered Blackbird (E)

The rediscovered Puerto Rican Nightjar and the recently discovered Elfin Woods Warbler are very local, though the other endemics are fairly common. Puerto Rican Flycatcher occurs on the Virgin Islands as well as on Puerto Rico.

Jamaica – 28 single-island endemics

Ring-tailed Pigeon (CE)
Crested Quail-dove
Yellow-billed Parrot
Black-billed Parrot (VU)
Jamaican Lizard-cuckoo
Chestnut-bellied Cuckoo
Jamaican Owl
Jamaican Mango
Black-billed Streamertail
Red-billed Streamertail
Jamaican Tody
Jamaican Woodpecker
Jamaican Elaenia
Jamaican Pewee
Sad Flycatcher
Rufous-tailed Flycatcher
Jamaican Becard
Jamaican Crow
White-eyed Thrush
White-chinned Thrush
Jamaican Vireo
Blue Mountain Vireo (NT)
Arrow-headed Warbler
Jamaican Euphonia
Jamaican Stripe-headed Tanager
Yellow-shouldered Grassquit
Orangequit
Jamaican Blackbird (NT)

The Jamaican Blackbird is threatened by habitat destruction. The other endemics are fairly easy to locate, but the Black-billed Streamertail is localised on the far eastern part of the island. Jamaican Oriole occurs also on the Cayman Islands.

Hispaniola – 27 single-island endemics

Ridgway's Hawk (E)
Hispaniolan Parakeet (VU)
Hispaniolan Parrot (VU)
Hispaniolan Lizard-cuckoo
Bay-breasted Cuckoo (VU)
Ashy-faced Owl
Greater Antillean Nightjar
Hispaniolan Emerald
Hispaniolan Trogon (VU)
Broad-billed Tody
Narrow-billed Tody (NT)
Antillean Piculet (NT)
Hispaniolan Woodpecker
Hispaniolan Pewee
Hispaniolan Palm Crow (NT)
White-necked Crow (VU)

La Selle Thrush (VU)
Flat-billed Vireo
Green-tailed Ground Warbler
White-winged Warbler (VU)
Hispaniolan Stripe-headed
    Tanager
Black-crowned Palm Tanager

Grey-crowned Palm Tanager
Eastern (Lowland) Chat
    Tanager (VU)
Palm Chat
Western (Highland) Chat
    Tanager (VU)
Antillean Siskin

Haiti, which forms the western part of Hispaniola, has Grey-crowned Palm Tanager as its only true endemic bird. The country is almost entirely deforested, and forest vegetation has been replaced with acacias; the lowland is barren. Native forest still occurs in steep ravines, in narrow strips along cliff bottoms and in very high mountains. Most endemic species with the exception of Grey-crowned Palm Tanager and Palm Chat are uncommon at best. The most endangered species of Haiti is White-winged Warbler. La Selle Thrush is locally common in the higher mountains.

In the Dominican Republic, Ridgway's Hawk is endangered, and White-winged Warbler and La Selle Thrush are vulnerable due to habitat destruction.

Cuba – 24 single-island endemics

Cuban Kite (CE)
Gundlach's Hawk (E)
Zapata Rail (E)
Blue-headed Quail-dove (E)
Cuban Parakeet (E)
Bare-legged Owl
Cuban Pygmy-owl
Cuban Nightjar
Cuban Trogon
Cuban Tody
Cuban Green Woodpecker
Fernandina's Flicker (E)
Giant Kingbird[1] (E)

Cuban Vireo
Cuban Palm Crow (E)
Cuban Crow
Zapata Wren (CE)
Cuban Gnatcatcher
Cuban Solitaire
Yellow-headed Warbler
Oriente Warbler
Cuban Grassquit[2]
Red-shouldered Blackbird
Tawny-shouldered Blackbird[3]
Red-shouldered Blackbird
Cuban Blackbird

Notes:
[1] Can now be considered a Cuban endemic as it is extinct in the other parts of its former range.
[2] Also on New Providence in the Bahamas where it was accidentally introduced in 1967.
[3] Scarce at the mouth of the Arbonite River in Haiti.
As well as these single-island endemic species, Cuba also shares many multi-island endemics with the Bahamas and the Cayman Islands.

Lesser Antilles
Many of the individual islands of the Lesser Antilles have their own endemic species:

Montserrat	Montserrat Oriole (VU)	
Guadeloupe	Guadeloupe Woodpecker	
Martinique	Martinique Oriole (E)	
Dominica	Red-necked Parrot (VU)	Imperial Parrot (VU)

St Vincent	St Vincent Parrot (VU)	Whistling Warbler (VU)
Grenada	Grenada Dove (CE)	
St Lucia	St Lucia Parrot (VU)	St Lucia Black Finch (NT)
	St Lucia Pewee	St Lucia Oriole (NT)
	Semper's Warbler (CE)	

(The St Lucia Nightjar has recently lost its status as a full species and is now a considered race of the Rufous Nightjar.)

Other West Indian endemics

Cayman Islands	Vitelline Warbler
San Andreas	San Andreas Vireo

(The San Andreas Mockingbird was formerly classed as a separate species, but is now considered to be a race of the Tropical Mockingbird.)

Multi-island endemics

These 65 species are endemic to the West Indies as a whole, but occur on two islands or more.

Black-capped Petrel (CE)
West Indian Whistling-Duck (VU)
Scaly-naped Pigeon
White-crowned Pigeon
Plain Pigeon (E)
Zenaida Dove
Caribbean Dove
Key West Quail-Dove
Bridled Quail-Dove (NT)
Grey-headed Quail-Dove (NT)
Rose-throated Parrot
Great Lizard-Cuckoo
Antillean Nighthawk
Lesser Antillean Swift
Antillean Palm Swift
Puerto Rican Emerald
Blue-headed Hummingbird
Antillean Mango
Purple-throated Carib
Green-throated Carib
Antillean Creasted Hummingbird
Vervain Hummingbird
Cuban Red-bellied Woodpecker
Antillean Mango
Purple-throated Carib
Green-throated Carib
Greater Antillean Elaenia
Crescent-eyed Pewee
Hispaniolan Pewee
Lesser Antillean Pewee
Grenada Flycatcher
La Sagra Flycatcher
Stolid Flycatcher

Puerto Rican Flycatcher
Lesser Antillean Flycatcher
Loggerhead Kingbird
Thick-billed Vireo
Black-whiskered Vireo
Cuban Crow
Caribbean Martin
Golden Swallow (NT)
Bahama Swallow (NT)
Red-legged Thrush
Forest Thrush(NT)
Rufous-throated Solitaire
Bahama Mockingbird
Scaly-breasted Thrasher
Pearly-eyed Thrasher
Brown Trembler
Grey Trembler
White-breasted Thrasher (E)
Adelaide's Warbler
Olive-capped Warbler
Plumbeous Warbler
Antillean Euphonia
Western Stripe-headed Tanager
Lesser Antillean Tanager
Cuban Grassquit
Greater Antillean Grackle
Jamaican Oriole
Tawny-shouldered Blackbird
Cuban Bullfinch
Greater Antillean Bullfinch
Lesser Antillean Bullfinch
Lesser Antillean Saltator

WHERE TO FIND BIRDS

Bahamas
Grand Bahama
Rand Nature Center, Freeport
Groves Botanical Garden,
    Freeport
Shannon Golf Course, Freeport

New Providence
Ardastra Botanical Gardens,
    Nassau
The Retreat (Headquarters of
    the Bahamas National Trust),
    Nassau
Versailles Gardens, Paradise
    Island

Abaco
Great Abaco Beach Hotel grounds
Crockett Drive
Sugarland Farm
Abaco National Park (south of
    Marsh Harbor)

The other islands of the Bahamas
are rarely visited by birders.

Puerto Rico
Caribbean National Forest ('El
    Yunque')
Humacao National Wildlife
    Refuge
Maricao State Forest
Guanica State Forest and Playa La
    Parguera area

Jamaica
Windsor Caves
Rockland Bird Sanctuary near
    Rockland
Marshall's Pen, near Mandeville
Mockingbird Hill Hotel grounds
Hardwar Gap

Dominican Republic
National Botanical Garden, Santa
    Domingo
Barahona and environs
Duverge, Puerto Escondido, and
    Lake Enriquillo
Aguacate
Las Mercedes
Los Haitesis National Park

Cuba
Bay of Pigs and Zapata Swamp
Cayo Coco
La Guira National Park

Lesser Antilles
Grenada
Hartman Estate

St Vincent
Vermont Nature trail in
    Buccament Forest
Dominica
Syndicate Nature Trail
Emerald Pool

Martinique
Le Jardin de Balata (botanical
    garden)
Reserve Naturelle de la Caravelle
St Lucia
Edmund Forest
Ravine de Chaloupe

Guadeloupe
Rio Corossol and picnic site on
    Basse Terre
Montserrat
Fogerty-Mars Hill

Other islands
Cayman Islands
South Sound Swamp
Little Cayman

San Andreas
All land habitats

MAP OF THE WEST INDIES

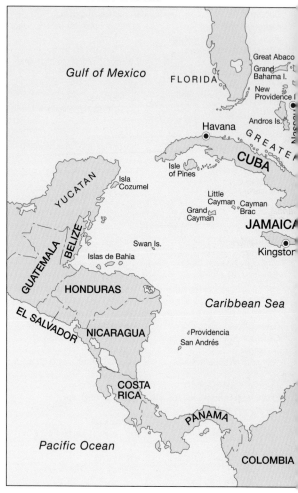

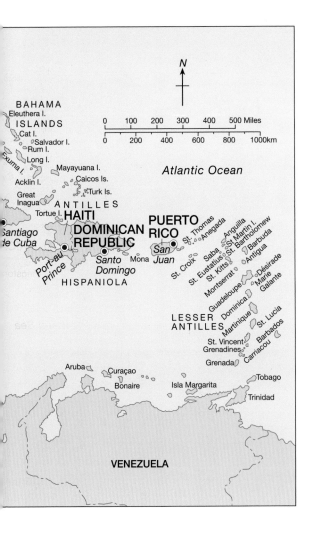

N

BAHAMA
Eleuthera I.
ISLANDS
Cat I.
Salvador I.
Rum I.
Long I.
Exuma I.
Mayayuana I.
Acklin I.
Caicos Is.
Great
Inagua
Turk Is.
Tortue I. **ANTILLES**
HAITI
DOMINICAN
Santiago **REPUBLIC**
de Cuba
Port-au
Prince
Santo
Domingo
HISPANIOLA

0   100   200   300   400   500 Miles
0   200   400   600   800   1000km

Atlantic Ocean

PUERTO
RICO
San
Juan
Mona
St. Thomas
Anegada
Anguilla
St. Martin I.
St. Bartholomew
St. Croix
Saba
Barbuda
St. Eustatius
St. Kitts
Antigua
St. Martin I.
Montserrat
Désirade
Guadeloupe
Marie
Galante
Dominica
LESSER
ANTILLES
Martinique
St. Lucia
St. Vincent
Barbados
Grenadines
Carriacou
Grenada

Aruba
Curaçao
Bonaire
Isla Margarita
Tobago
Trinidad

VENEZUELA

PIED-BILLED GREBE *Podilymbus podiceps* 30–38cm

This is the common grebe of the West Indies, resident on ponds and lagoons. Its bill is larger and more conical than in Least Grebe; adults are greyish-brown, with a black band on the bill. It is most common on the larger islands, least common on smaller ones. Usually observed in pairs or small groups in fresh, or less often brackish water. It feeds by diving, taking a wide variety of aquatic animal fare, and will disappear quickly under the water if it is disturbed.

WHITE-TAILED TROPICBIRD *Phaethon lepturus* 91–107cm

B. Hallett

The 'bosun bird' is a spectacular species. It is snow-white, the heavy bill is yellow to orange, and the central tail plumes are longer than its body; there are heavy black bars on the upperwings which show in flight. A local breeder throughout the West Indies, it is uncommon in the Lesser Antilles, where displaced by Red-billed Tropicbird. Usually observed soaring over sea cliffs where it nests. It is a pelagic species which disperses up to 1,000 km after the breeding season. Feeds on flying fish and squid by plunge-diving.

RED-BILLED TROPICBIRD *Phaethon aethereus* 90–106cm

R. Behrstock; Academy of Natural Sciences

The Red-billed Tropicbird is a common breeder in the Virgin Islands and in the Lesser Antilles and is partial to small islands when it comes to choosing nesting sites. It is the heaviest of the tropicbirds. The plumage is primarily white, with a black streak through the eye, and the upperparts are barred with black; the bill is bright red. In flight there is a black band on the outer wing edge. Feeds on flying fish and squid.

RED-FOOTED BOOBY *Sula sula* 66–76cm

There are two colour phases, the brown phase illustrated here; note white hindparts. The other phase is entirely white with black primaries and secondaries. Both of the phases have bright red feet. This gregarious bird nests on remote islands, and a large colony of over 18,000 is present on Little Cayman. It is the only booby which nests in trees and bushes. Groups of boobies fly in V formation when coming to roost from their fishing grounds at sea, sometimes hundreds of miles from land. Feeds mainly on flying-fish.

BROWN PELICAN *Pelecanus occidentalis* 107–137cm

Easily identified by its large size, brown colour and huge bill. Common in the Bahamas, Greater Antilles and northern Lesser Antilles; uncommon elsewhere. It breeds in colonies, but is most often seen around fishing boats, docks and piers searching for scraps and handouts. It can dive from great heights, its entire body disappearing below the surface. It flies in small groups with synchronous wing-beats, gliding for long stretches. A coastal species, avoiding both the open sea and fresh water. Roosts in mangroves.

NEOTROPIC CORMORANT
Phalacrocorax brasilianus 63–69cm

The Neotropic Cormorant is most fond of fresh water. It is black overall; its throat pouched is lined with white. Less bulky than the Double-crested Cormorant; it has a thinner beak and longer tail. Breeds in Cuba and the Bahamas, and appears to be expanding its range eastward through the West Indies. Previously known as the Olivaceous Cormorant. Feeds on a greater variety of aquatic life than the Double-crested; food includes crustaceans, frogs, aquatic insects and fish, which it pursues under-water.

DOUBLE-CRESTED CORMORANT
Phalacrocorax auritus 74–89cm

Characteristic features are the dark overall colouration, large size, long neck, hooked bill and habit of perching with wings spread to dry them. It is larger than Neotropic Cormorant. Though chiefly a saltwater species, it does occur in fresh water as well. It is resident in Cuba and the Bahamas (San Salvador), and a wanderer elsewhere in the West Indies. It swims and dives well, feeding entirely on fish taken in open water. It can sink slowly out of sight if alarmed. Flocks fly in V formations, usually low over the water.

Adult (above); immature (below)

Female

Male

ANHINGA *Anhinga anhinga* 85cm

A freshwater species with a long, thin neck and a long dagger-like bill to impale fish under water. The Anhinga has a large whitish patch on the back and upperwings, and frequents shallow, calm bodies of water. Like the cormorants – to which it is closely related – it dries its wings intermittently in the sun. It often swims below the surface with only the head and neck showing hence the name 'snake-bird'. Frequently soars with neck extended and tail fanned. Regular in Cuba only, and a vagrant elsewhere in the West Indies.

MAGNIFICENT FRIGATEBIRD *Fregeta magnificens* 94–104cm

The 'man-of-war' bird has a long, forked tail; it soars very slowly on long pointed wings, in and around coastal areas. Males are entirely black with a red throat in the breeding season. Females are white-breasted, and the immature sports a white head and breast. Frigatebirds are the hawks of the sea, relying on speed and agility to gather food from the ocean surface, especially flying-fish and squid, or to rob other species. This is the only species of frigatebird which occurs in the West Indies, where it breeds throughout.

GREAT BLUE HERON *Ardea herodias* 107–132cm

Normal   *White phase*

This, the largest heron of the Americas, is common in the West Indies. An uncommon white form occurs in the Caribbean, considerably larger than the otherwise-similar Great Egret; it also differs from that species in having yellow legs. The food is a wide variety of prey such as mice, crabs, shorebirds and frogs, as well as fish. Birds nest singly or in colonies, the size of which is dictated by the presence of suitable large trees which are isolated from disturbance.

GREAT EGRET *Egretta alba* 89–107cm

The Great Egret is common in the West Indies, but uncommon in much of the Lesser Antilles. This large, entirely white, long necked species has beautiful plumes in breeding plumage. The bill is long and yellow while the legs are blackish. A dark line extending from the gape to beyond the eye occurs in no other white heron. The Great Egret lives and feeds on aquatic life in fresh and saltwater marshes, swamps, and shallows of lakes. Feeds on fish, frogs, other aquatic prey and even mice.

SNOWY EGRET *Egretta thula* 51–71cm

A beautiful species which can well be described as 'dainty'. Its breeding plumage is bright white and very conspicuous. Legs are black, the feet bright yellow to orange and the lore bright yellow. The bill is black and appears pencil thin, tapering to a fine point. Snowy Egret is common in the West Indies, but patchily so in the Lesser Antilles. An active feeder. Though sometimes confused with the immature Little Blue Heron, that species is a deliberate feeder, and has greenish legs and a thicker bill; immature Little Blue Herons have no head plumes.

LITTLE BLUE HERON *Egretta caerulea* 56–71cm

Adult   Immature

The adult Little Blue Heron is easily identified by its dark grey colouration, but the white immature birds can be mistaken for a number of species. Feeds on a diversity of aquatic fare. Moves freely between Cuba and Yucatan. The young Little Blue Heron has no plumes, the legs are greenish overall. Though present commonly in the West Indies throughout the year, it does not breed there. When water dries up in wetlands, it adapts to feeding on grasslands.

TRICOLORED HERON *Egretta tricolor* 61–71cm

This heron is uniformly grey above with a white belly and under tail-coverts; it is slightly taller than Little Blue Heron, and appears small and slim due to its long slender neck. In the West Indies it occurs mainly in mangroves and saltwater lagoons; sometimes fresh water near the coast. It is a common bird through much of the West Indies, though rare in the Lesser Antilles. An acrobatic feeder, its normal fare is aquatic insects, small fish, frogs and tadpoles, as well as other small aquatic life. It wades in thigh-deep water, and is a solitary feeder.

REDDISH EGRET *Egretta rufescens* 69–81cm

Intermediate in size between Great Egret and the medium-sized herons. There are two distinct colour forms. A dark form which is greyish overall, with the head and neck reddish-brown, and a white form which is totally white. The best field mark for this species, along with its size, is the black-tipped bill, which is pinkish at the base. Reddish Egret appears to have a generally ruffled appearance, and its habit of dancing around in shallow water pursuing its prey is characteristic. It is a common year-round resident in the Bahamas and Cuba, but is uncommon to rare elsewhere.

CATTLE EGRET *Bubulcus ibis* 48–64cm

Non-breeding

An egret that feeds in open country and pastures on insects and other creatures disturbed by cattle – hence its name. The population of the Cattle Egret is exploding, and its numbers have overwhelmed those of native herons. It soon takes over the colonies of other herons, where it becomes the major occupant. This yellow-billed heron with its hunched posture is orange-buff on the neck, chest and back during the breeding season, though as a non-breeder its plumage is pure white and the legs are dark.

GREEN HERON *Butorides virescens* 40–48cm

Adult   Immature

This species is common throughout the West Indies. Its small size, variable but always dark colouration, short chestnut-coloured neck, and short, yellow to orange legs are distinctive. A rare rufous form is confined to Cuba. The immature is brownish above, and whitish with heavy brown streaks on the underparts. It is usually a solitary bird, and a skulking one, and can be found by waterbodies of any size.

BLACK-CROWNED NIGHT HERON
Nycticorax nycticorax 58–71cm

The adult of this medium-sized, compact heron has a black crown and back, with a black-and-white face and underparts. It sports head plumes. The immature is quite different, being brown with white specks. In flight the legs extend beyond the tail. It is a common resident in the West Indies where it breeds on the larger islands. Nocturnal for the most part, it can be seen most often at dawn and dusk flying between its roost and the feeding areas. It lives and feeds in all types of water, whether fresh, brackish or salt.

YELLOW-CROWNED NIGHT HERON
Nyctanassa violacea 56–71cm

Joel Greenberg

The Yellow-crowned Night Heron is a medium-sized but rather chunky grey heron. The adult is dark grey with black and yellowish-white head markings. The immature resembles the young Black-crowned Night Heron but is light grey with finer specks. In flight the legs of this species extend well beyond the tail. Solitary and nocturnal, it feeds on hard-shelled invertebrates for the most part. Feeds and breeds in a variety of habitats, though preferring saltwater areas.

WHITE IBIS *Eudocimus albus* 56–71cm

The adult is unmistakable, being pure white with a bright red face and legs, and a long sickle-shaped bill. It shows some black on the tips of the primaries. The immature is dark brown above, and white below and on the rump, but is easily identified by the diagnostic bill. Common on Cuba and Hispaniola, local on Jamaica and rare elsewhere. Feeds on aquatic life which it probes for in the mud. The closely related Scarlet Ibis *E. ruber* of South America is a vagrant to the West Indies.

GLOSSY IBIS *Plegadis falcinellus* 56–64cm

The Glossy Ibis is a highly nomadic wanderer which undertakes intermittent migrations into the West Indies. Thus it is generally sparse in occurrence, breeding only irregularly in the Greater Antilles, and visiting the US Virgin Islands and the more northerly of the Lesser Antilles. Its overall dark-looking colouration – showing as iridescent violet and green only at close range or in good light – and its long, thin, downcurved bill are diagnostic. The food is insects, crayfish and even small snakes.

ROSEATE SPOONBILL *Ajaia ajaja* 66–81cm

The Roseate Spoonbill is aptly named. The bare greenish head, long spoon-shaped bill and carmine wing coverts, which shade into pale pink, are distinctive. The legs are red. White immatures have pale pink wings, with yellow legs and bill. In the West Indies it is an uncommon resident on Cuba and Hispaniola; vagrant in Jamaica; local in Puerto Rico; uncommon to rare elsewhere. Feeds by sweeping its bill from side to side in shallow water, primarily catching fish, but invertebrates and plant material are also important, and are sifted from the mud.

WOOD STORK *Mycteria americana* 100cm

In the West Indies the Wood Stork breeds only locally in Cuba, and is a vagrant elsewhere. The large size, unfeathered dark head, long and slightly downcurved bill, and white plumage make it an easy bird to identify, even at a distance. It is commonly observed soaring, when the black tail and trailing edges to the wing can be seen; the feet extend beyond the tail. Feeding by touch and taking all types of aquatic animal matter, it occurs in swamps, mangroves, mudflats, wet meadows and man-made bodies of water.

TURKEY VULTURE *Cathartes aura* 68–80cm

Large size, bare reddish head and black and silvery-grey underwing are diagnostic. It soars effortlessly with wings raised in a shallow V, seldom flapping them. Common in the Bahamas and Greater Antilles, yet absent from the Lesser Antilles. With highly developed vision and olfactory senses, it is nature's sanitarian, feeding on carrion, especially road kills; occurs in great numbers at dumps and landfills. Assembles in large roosts, and at dawn the bird spreads its wings to absorb solar heat. Sometimes soars in numbers, especially on cloudy days.

GREATER FLAMINGO *Phoenicopterus ruber* 107–122cm

Unmistakable even at great distance, this gregarious species first appears as a sea of pink, which can number in the thousands. Birds filter-feed with heads inverted, straining plankton and invertebrates from the silt. Nesting is in shallow saltwater bays, and one egg is laid atop a cone of mud. Flamingos are good swimmers; they simply fold their legs and paddle with their webbed feet. Scattered about throughout the West Indies, but rare and local. The largest colonies found at Great Inagua in the Bahamas, and in Cuba; there is a small colony in the Dominican Republic.

WEST INDIAN WHISTLING-DUCK
Dendrocygna arborea 48–56cm

The largest of the whistling ducks; it flies with neck extended giving it a humpbacked appearance; the legs extend well beyond the tail, making this species appear ungainly in flight. The West Indian Whistling Duck is endangered throughout its range. It is an uncommon resident in the Bahamas, Cuba, Jamaica, Puerto Rico, Dominican Republic and Cayman Islands, and there are also populations in the Virgin Islands and Antigua. It roosts in swamps and mangroves during the day and feeds at night in agricultural fields. Often seen perched in trees.

MALLARD *Anas platyrhynchos* 51–71cm

The Mallard is perhaps the most well-known of ducks but is only a rare non-breeding resident or vagrant in the West Indies. The green head, bright yellow bill, white neck-ring and brown breast of the male are distinctive; the female is warm brown with an orange bill. The species is highly adaptable, birds being seen commonly in city parks and on almost any body of calmer water. The Mallard feeds on a great diversity of vegetation including seeds, plants and submerged roots, along with aquatic invertebrates.

BLUE-WINGED TEAL *Anas discors* 38–40cm

The Blue-winged Teal is the most common duck to be found in the West Indies, and is present all year, occurring on virtually all islands – yet it does not breed in the region. The drake has a distinctive bold white crescent face-mark, which curves in front of the eye, while the female is warm buff spotted with dark brown. In flight both sexes show a paddle-shaped, pale blue inner forewing. Feeds on vegetation in almost all types of waterbody.

WHITE-CHEEKED PINTAIL *Anas bahamensis* 38–48cm

The White-cheeked Pintail – also known as the Bahama Pintail – is easily identified by its red bill, contrasting white cheek and its overall reddish-brown colouration. It is resident throughout the West Indies, but commutes between islands when necessary. It is a declining species which is threatened by overhunting and by destruction of suitable habitat. It lives generally in mangrove swamps, and on pools of brackish or salt water; it is uncommon in fresh water.

NORTHERN PINTAIL *Anas acuta* 51–56cm

A distinctive and elegant duck. Both sexes sport a long pointed tail (longer in the male) which is visible both in flight and on the water. The drake has a chocolate-brown head, neck and hindneck, and a white breast extending to and narrowing along the neck; underparts are grey, and the ventral region buffy and black. The female is fawn-brown. It is an uncommon non-breeder in the Greater Antilles (except Jamaica) and scarce elsewhere. A freshwater species which occasionally inhabits coastal saltwater lagoons. Feeds by upending to eat marsh vegetation.

LESSER SCAUP *Aythya affinis* 38–46cm

The 'Bluebill' drake is grey-backed with a dark purple head and breast. It can be confused initially with Ring-necked Duck, which has a black back. The brown female's flanks are barred, the face white. It is similar to Greater Scaup *A. marila* which is a rare bird in the West Indies. Lesser Scaup in flight shows a white wing stripe confined to the secondaries only, whereas the Greater's extends to the wing-tip. A local non-breeding resident in Cuba and the Bahamas; uncommon to rare elsewhere. Frequents coastal lagoons, lakes and estuaries, but also large bodies of fresh water.

RED-BREASTED MERGANSER *Mergus serrator* 51–64cm

The drake Red-breasted Merganser is a ragged-crested, green-headed duck which has a long thin serrated bill, a white collar and back and a reddish breast. The female also has a rough crest on the rear of its head, but the head is brown and the body grey. A rare winter resident in the Bahamas and in the Greater Antilles, especially Cuba. It lives on open areas of coastal sea, especially bays, and on nearby lagoons, where it dives to catch fish.

RUDDY DUCK *Oxyura jamaicensis* 35–43cm

A small diving duck, the drake of which is chestnut with a dark crown and nape, a white face and a pale blue bill. Its dark brown tail is usually cocked upright. The female is pale brown with a brown cap with a single stripe across the face. A locally common breeding species in New Providence and the Greater Antilles, but uncommon to rare elsewhere. Frequents wetlands which range from deep bodies of fresh water to brackish lagoons with emergent vegetation. Feeds on vegetation sifted from bottom mud, as well as miscellaneous animal material.

OSPREY *Pandion haliaetus* 53–61cm

The Osprey's simple requirements are water and fish. Eagle size, the head and neck are white. There are black patches on the head and nape. The underparts are white, but may be lightly streaked. While soaring, the Osprey kinks the wings at the wrist, rendering it unmistakable in flight. In the West Indies, the resident race is common throughout, but breeds only in the Bahamas and Cuba. Nests are huge, and are added to yearly. The local race is entirely white-headed, and completely white below.

SHARP-SHINNED HAWK *Accipiter striatus* 25–35cm

The Sharp-shinned Hawk is a small forest hawk whose wings are short and rounded. Its head appears small, and it has a long, narrow, square tail which is barred with black. Except for the endemic Gundlach's Hawk of Cuba, it is the only native accipiter in the West Indies. The Puerto Rican race is pictured. This species is found throughout the Americas. The West Indian residents are in Cuba, Hispaniola and Puerto Rico; some North American birds migrate to the Bahamas, and to other islands during the winter. It occurs in mature forests at higher altitudes. It feeds mainly on small birds.

Kevin T. Karlson

COMMON BLACK HAWK *Buteogallus anthracinus* 51–58cm

An unmistakable chunky, sooty-brown raptor, with a single wide, white tail-band and white patches at the base of the primaries, which is common along the Cuban coast. Here it perches on fence posts and trees along bays and estuaries, and feeds on crabs, amphibians and crayfish. This Cuban form is sometimes considered a separate species, as it has different calls and larger feet compared with continental birds. The species occurs also on St Vincent, though it is uncommon and frequents mountain forest. It has a typical flapping and gliding flight.

BROAD-WINGED HAWK *Buteo platypterus* 35–41cm

Immature   *Sub-adult*

This is a year round resident of the West Indies, where it is joined by some North American migrants. Common in Cuba, and the Lesser Antilles, a vagrant elsewhere. The West Indian form is smaller and redder. Small and chunky, brown above, white below with the tail widely banded black and white. The Broad-winged and Red-tailed hawks are the common buteos of the West Indies. Broadwings range from North through Central and northern South America. Soars frequently, feeds upon insects, reptiles and small birds in forests of all elevations. North American birds are highly migratory.

RED-TAILED HAWK *Buteo jamaicensis* 48–64cm

The Red-tailed Hawk is the most common species of *Buteo* in the West Indies. A large hawk which is identified by its broad rounded wings and reddish tail. It is white below with a black breast-band. The immature has a brownish tail with heavier streaking on the underparts. It is a resident on the larger islands of the Bahamas, the Greater Antilles, northern Lesser Antilles and uncommon to rare elsewhere. Feeds on introduced rodents, lizards, snakes and large invertebrates. It is widespread in most habitats, sometimes within urban lowland areas.

CRESTED CARACARA *Caracara plancus* 50–63cm

The Crested Caracara is quite distinctive whether perched or in flight. It has a large crested head, a large beak and red facial skin; the breast is barred black and white. In flight large white windows are evident near the wing tips. The immature is browner, but the head and beak are distinctive. Local and widespread on Cuba, rare elsewhere. Ranges from the southwestern United States through South America. Feeds on birds, lizards and rodents; sometimes observed with vultures feeding on carrion. Voice is a cackling call.

AMERICAN KESTREL *Falco sparverius* 23–30cm

Male   Female

The American Kestrel is a common year-round resident in the West Indies ranging south to St Lucia. It is a small and colourful falcon, the male being generally reddish-brown on the back and blue-grey on the wings; the tail is reddish-brown with a black band near the tip; there are spots of reddish colour on the head and two short vertical black bars on the side of the face. The female is similar but has brown wings instead of the male's grey. Feeds in open areas on mice, small birds, insects and lizards.

CLAPPER RAIL *Rallus longirostris* 36cm

The Clapper Rail is a bird of the salt and brackish water marshes. A grey bird with long legs and beak, observed within the mangroves or salt marsh is almost certainly this species. In the West Indies it is a permanent resident in the Greater Antilles, where it is common in Puerto Rico, Cuba and the Bahamas; local elsewhere, also Guadalupe, Martinique and St. Christopher. Feeds upon wide diversity of aquatic invertebrates, and vegetation which it seizes by probing into the silt. Can swim and dive well. When in flight it lets its legs dangle as it drops into nearby vegetation. If disturbed it wades and runs through the shallows and over exposed roots. Easy to observe at low tide.

SORA *Porzana carolina* 22cm

The Sora can be found throughout the West Indies and at all times of year, though it does not breed in the region. It occurs commonly in the Bahamas and Cuba, but is less often encountered elsewhere. It is easily distinguished by its brownish-grey colouration, black face, and stubby, bright yellow-green bill. Usually the easiest of the rails to see, as it inhabits mangroves, marshes, rice fields and other fresh, brackish or saltwater areas. Feeds on seeds of water plants and invertebrates.

PURPLE GALLINULE *Porphyrula martinica* 27–36cm

This colourful species has bright bluish-purple plumage, a bluish-white frontal shield on the forehead and bright yellow legs with long toes. Common in Cuba and Hispaniola, uncommon in Puerto Rico, Jamaica, the Caymans and San Andreas; it is a migrant in the Bahamas and rare elsewhere in the West Indies. Lives in freshwater marshes, ricefields, and fringes of ponds, channels and other watercourses. Climbs to the top of taller swamp vegetation, where it roosts during the evening. Feeds on a large variety of vegetation and aquatic animal life.

COMMON MOORHEN *Gallinula chloropus* 30–38cm

The yellow-tipped red bill, red frontal shield and white line along the flank distinguish this species. For the most part it is a grey-brown duck-like bird with long legs and toes. Common throughout the West Indies as a permanent resident. Lives in freshwater wetlands with emergent vegetation, and either still or running water. It swims readily, with a bobbing head motion, and is usually rather tame. An opportunistic feeder, feeding on a wide range of animal material and vegetation. Diurnal, but active on moonlit nights.

AMERICAN COOT *Fulica americana* 34–43cm

An unmistakable bird, generally grey-black with a white bill and frontal shield. It has a smaller head than Purple Gallinule and Common Moorhen, and unlike them it is generally seen in open water where grebes and other waterfowl are feeding. Feeds on submerged vegetation by diving and dabbling, but will graze in open fields. Its long toes are edged with fleshy lobes and it is a strong swimmer, though also walking well. Sometimes observed in large flocks. Common throughout the West Indies, but uncommon as a breeder.

LIMPKIN *Aramus guarauna* 69cm

The Limpkin is a large bird with long legs and a bill which is slightly downcurved; its general colouration is brown with bold white spotting. Long toes enable it to navigate aquatic vegetation. Feeds for the most part on large aquatic snails, but also crustaceans, frogs, lizards and insects. In the West Indies it is common only on the northern Bahamas and Cuba; local in Hispaniola and Jamaica; and absent from Puerto Rico southward. Known as the 'crying bird'; it is usually tame when encountered.

SANDHILL CRANE *Grus canadensis* 88–95cm

A North American species which is also resident in Cuba. The Cuban form is non-migratory, and frequents marshes, swamp edges, and open or semi-open areas including the edges of pine barrens and savannas. This all grey species is considerably larger than the Great Blue Heron, and shows a bare red crown. It flies with neck fully extended; its call is a loud bugle. Pairs dance and bugle together during the breeding season. Forages in small groups when not breeding. Monogamous and shy.

DOUBLE-STRIPED THICK-KNEE *Burhinus bistriatus* 38–48cm

This large-headed, long-legged, plover-shaped species occurs in the West Indies only locally on Hispaniola. The large iris is yellow, there is a white stripe over the eye, and large white wing-patches are displayed in flight. The upperparts are brown; the breast is streaked, and the rest of the underparts are white. A ground-dweller, it frequents arid open areas, savannas and cultivated areas where it feeds chiefly at night, on insects, lizards, small mammals and other animal fare. Takes flight reluctantly; usually occurs in pairs.

BLACK-BELLIED PLOVER *Pluvialis squatarola* 26–34cm

This is the largest of American plovers, and also known as Grey Plover. Its stocky build, short bill, greyish-white general colouration and – when in breeding plumage – its black underparts serve to identify it. Usually seen feeding along the water's edge, mainly on or near the coast. It is a species which is found worldwide, and is a common sight throughout the West Indies at all times of year as a migrant or winter visitor.

WILSON'S PLOVER *Charadrius wilsonia* 18–20cm

B. Hallett

The Wilson's or Thick-billed Plover is brown above, white below; it has a large thick, black beak. The male sports a large, broad black breast band ; that of the female is brown. Lives on coastal beaches, and mudflats. Unlike other small plovers, prefers to escape danger by running, so it tends to remain well away from vegetation and areas covered by flotsam. Feeds on a great diversity of aquatic life as it forages along the edge where waves cast up a better variety. A common permanent resident of the Bahamas and Greater Antilles; rarer in the Lesser Antilles. Occurs in warmer areas of the western hemisphere.

KILLDEER *Charadrius vociferus* 25cm

The Killdeer is the most recognised shorebird within its range. A large plover which is brown above, and white below with a pair of black breast bands, and a reddish brown rump which flashes in flight. The shrill, loud, monotonous call resembles its name. It lives in open habitats usually in agricultural land, savannas, and other grassy and wetland areas, as well as estuaries. Frequently encountered on parking lots and construction sites. The West Indian race is a bit smaller and greyer but is generally indistinguishable in the field. In the West Indies it is common in the Greater Antilles, the Bahamas, and the Virgin Islands; uncommon elsewhere.

AMERICAN OYSTERCATCHER *Haematopus palliatus* 43–46cm

This bright and flashy black and white shorebird inhabits rocky coasts where it feeds primarily on shellfish, crabs, small starfish, and other salt water fare; its beak is used to expose shellfish by prying or pounding. The selection of this feeding technique is based upon the method taught by the parent. This crow size species has a bright red laterally flattened beak; the eyes are orange, circled with red lids. The legs are flesh coloured. A common, but local species in the West Indies in the Bahamas, Puerto Rico and the Virgin Islands; uncommon elsewhere. It can swim and dive well when necessary.

BLACK-NECKED STILT *Himantopus mexicanus* 34–40cm

Only the flamingo has longer legs in proportion to its size. This delicate, unmistakable, black-and-white shorebird is endowed with bright red legs and a slender bill. In flight the long slender wings are entirely black, the rump and tail pale grey. It is usually observed in flocks where it occurs in a great variety of aquatic habitats both fresh- and saltwater. The diet is small aquatic organisms, taken from dry mud or in belly-deep water. Common and widespread through most of the West Indies; scarce in the southern Lesser Antilles.

GREATER YELLOWLEGS *Tringa melanoleuca* 29–38cm

Very similar to Lesser Yellowlegs, but larger and with a longer, thicker bill which may be slightly upturned. It is a large sandpiper with bright yellow legs, and underparts which are speckled in spring, pure white in winter. The dark grey-brown of the back is speckled or streaked with white; wings are dark, the rump white. It is a non-breeder present all year throughout the West Indies, where it is common on mudflats and shallow wetlands both fresh and saline. Three- or four-note whistling call.

LESSER YELLOWLEGS *Tringa flavipes* 23–28cm

Similar to but considerably smaller than Greater Yellowlegs, though size is usually difficult to establish in the field unless both species happen to be observed together. Lesser Yellowlegs has a bill which is much the shorter, and their sizes do not overlap. Like Greater Yellowlegs it is brownish-grey above, streaked with white; the underparts are white and the legs bright yellow. Occurs commonly all year throughout the West Indies on mudflats and in shallows in fresh or salt water.

WILLET *Catoptrophorus semipalmatus* 33–41cm

Transition plumage *Winter*

A large grey shorebird which is light grey in overall colouration, and has long bluish-grey legs and a similarly coloured, medium-length and rather thick bill. In flight the black-and-white-striped wing pattern is very distinctive. It is a permanent resident of the Bahamas, Caymans and Greater Antilles, and occurs as a non-breeder elsewhere in the West Indies. Mud- and sandflats with saltmarsh vegetation in bays and estuaries along the coast are preferred.

SPOTTED SANDPIPER *Actitis macularia* 18–20cm

This common sandpiper has bright white underparts with black spots; there is a smudge on the side of the neck, the base of the beak is yellow; it becomes orange during breeding. Easily recognised by its stiff rapid wingbeats and white wing stripe. It teeters and bobs as it walks along. A common non-breeding resident throughout the West Indies, lacking spots in winter. It forages alone or in pairs along water edges both fresh and salt.

RUDDY TURNSTONE *Arenaria interpres* 21–25cm

The Ruddy Turnstone, a chunky, plover-like bird, is easily recognised by its habit of flipping shells and rocks to feed on aquatic hidden life beneath. The harlequin pattern of the breeding plumage is rarely to be seen in the West Indies, and the birds present are more usually distinguished by their dark breast markings and bright orange legs. In flight there is a distinctive white pattern on the back, tail and upperwing. Present all year in the West Indies, on mudflats and on sandy and rocky coasts.

Winter (top);
winter moulting (left)

SANDERLING *Calidris alba* 18–22cm

The Sanderling is an unmistakable bird of the beach, where it scampers along the shore and skirts the waves to snatch food items. Then quickly it retreats from an incoming breaker in the fashion of a wind-up toy. Its light grey upperparts and white underparts allow it to become virtually invisible as it races along on short black legs. In flight it displays a bright white wing-bar. It is common all year as a non-breeder throughout the West Indies.

WESTERN SANDPIPER *Calidris mauri* 14–18cm

The Western Sandpiper is very similar to the Semipalmated, but differs from it in having a longer, thicker bill, which usually droops slightly at the tip. The upperparts are brownish and the underparts are white. It usually feeds in deeper water than do Semipalmated and Least Sandpipers. More common along the coast than inland. Feeds on small invertebrates, mainly by probing. Forms flocks with other 'peeps' and is common all year as a non-breeder throughout the West Indies.

LEAST SANDPIPER *Calidris minutilla* 13–16cm

This tiny species, smallest of the sandpipers which occur in the West Indies, is, like the other 'peeps', brownish above with white underparts, but its yellow legs serve to differentiate it. Other useful features to be looked for are its short thin bill and a pale stripe along the side of the back. It forms flocks with Semipalmated Sandpipers on coastal mudflats and inland freshwater borders where it feeds on small invertebrates. A common migrant in the West Indies.

SHORT-BILLED DOWITCHER *Limnodromus griseus* 25–30cm

Easily recognized by its long straight bill which it probes into the mud, immersing it completely and pumping it vertically up and down in the manner of a sewing machine. This grey, chunky, medium-sized sandpiper has yellowish-green legs, and in flight it displays a prominent white patch on the back. Bays sheltered from the wind are the preferred feeding areas. The 'tu-tu' note serves to distinguish it from the Long-billed Dowitcher *L. scolopaceus* which is rare in the West Indies. The Short-billed Dowitcher is a common but local non-breeder in the West Indies.

LAUGHING GULL *Larus atricilla* 38–43cm

The common gull of the West Indies; it breeds locally, and occurs universally in coastal areas. The only West Indian gull which has a black head. Outside the breeding season the head is mottled black on the back and sides, but the white border on the trailing edge of the wing is distinctive in flight. The dark grey mantle and black wing-tips are also diagnostic. The immature is greyish-brown above with a dusky breast. The call is a laughing 'ka-ka-ka-ka-ka-kaa-kaa-kaaa'. Its numbers are increasing dramatically.

Winter (above); immature (below)

RING-BILLED GULL *Larus delawarensis* 46–51 cm

The Ring-billed Gull is not as common as the Laughing Gull in the West Indies. It is a non-breeder which is rare in the Lesser Antilles. The adult Ring-billed is larger, and has a yellow beak with a dark ring, and greenish legs. It is distinguished from the Laughing Gull by the absence of a black head and the beak and leg colour, which in the Laughing Gull are dark. Similar but smaller to the Herring Gull which is rare and local in the area. Feeds on fish and other aquatic life, and is resident at refuse dumps. Occurs from coastal areas and harbours to fields and parking lots.

GULL-BILLED TERN *Sterna nilotica* 35–38cm

Most terns have thin bills, short legs and a crest, but the Gull-billed Tern is well named for it has a broadly thickened bill, rather long legs and a rounded head. When flying, its uniform grey upperparts extending to the rump and tail serve to separate it from Sandwich Tern. The flight is gull-like, and birds do not dive for their prey but instead pick it up from the surface. Gull-billed Tern is an uncommon breeding resident in the Bahamas; uncommon on Puerto Rico and Hispaniola and winters elsewhere in the Caribbean.

ROYAL TERN *Sterna maxima* 45–53cm

The large size and bright orange-yellow bill of this species are diagnostic. The Royal Tern is as large as a medium-sized gull; the underside of the primaries are white with black trailing edges to the tips. It has a prominent crest at all seasons. A common coastal species throughout the West Indies, but local as a breeder. The only similar species likely to be encountered is the Caspian Tern *Sterna caspia*, which is a bit larger with a huge dark red bill; the Caspian is rare in the West Indies.

SANDWICH TERN *Sterna sandvicensis* 36–41cm

Kevin T. Karlson

Any crested tern having a long slender black bill with a bright yellow tip is most certainly this species. In South America the 'Cayenne Tern' which has an entirely yellow bill is thought to be only a southern race of the Sandwich Tern, and it occurs rarely but regularly in the West Indies. The upperparts are pale grey; the rump and tail white, whereas the upperparts of Gull-billed Tern are entirely pale grey. A permanent resident in the northern part of the West Indies, and after breeding it wanders southward.

LEAST TERN *Sterna antillarum* 21–24cm

Kevin T. Karlson

The Least Tern is the smallest tern species of the West Indies. It is a swift, buoyant flier with a black crown, a white forecrown, a yellow bill with a black tip, and yellow legs. It is a common resident in the northern part of the West Indies, but only a wanderer further to the south. A solitary breeder near the high-water mark on sandy islands. Feeds along coasts, and inland along rivers, plunge-diving for small aquatic items; occasionally insects are taken.

SOOTY TERN *Sterna fuscata* 33–43cm

B. Hallett

A large, black-and-white, tropical tern with a powerful flight, usually seen far from land. Soars and feeds in flocks, preferring to take fish from the surface rather than to dive. A white streak originates from its forehead, but extends only to a point above the eye, never behind it. Unlike Bridled Tern *S. anaethetus* there is no white across the hindneck. A deeply forked tail, white underwing-coverts and white outer tail-feathers complete its description. It is a common resident breeder on coral atolls throughout the West Indies.

BROWN NODDY *Anous stolidus* 38 cm

B. Hallett

The Brown Noddy nests on offshore cays, but is normally a pelagic species rarely seen from land. Its compact appearance, overall dark brown plumage, pale white cap and long wedge-shaped tail are distinctive. Sometimes occupies breeding areas with Sooty Terns. This species does not swim nor dive for its prey, but instead takes it from the surface. Common throughout the West Indies; occurs throughout the tropical and subtropical seas of the world. Feeds on small fish and squid, sometimes by moonlight.

SCALY-NAPED PIGEON *Columba squamosa* 36–40cm

The Scaly-naped (or Red-necked) Pigeon is a permanent resident throughout the West Indies, and otherwise occurs only on islands off Venezuela. This beautiful, large woodland and rain forest pigeon is an arboreal species which can be found occurring either singly or in small flocks. The general colouration is grey at a distance, but upon closer observation the head, neck and breast are purplish-red. The hindneck is chestnut and metallic purple. The feet are red; the eye-ring of the male is red, of the female yellow.

WHITE-CROWNED PIGEON *Columba leucocephala* 29–40cm

The White-crowned Pigeon is a common permanent resident of the West Indies. It moves freely among the islands where it prefers wooded areas at lower elevations. A very distinctive bird in flight, as the conspicuous white crown contrasts with the dark grey body, though in young birds and females the crown is rather greyer or browner. The hindneck is iridescent, but this is difficult to see in any but good light. Arboreal, it nests in mangroves, feeds on fruit and flowers, and on some islands it undergoes altitudinal migration.

PLAIN PIGEON *Columba inornata* 38–40cm

A large, pale purplish-brown pigeon with pale grey tail and rump and white-margined wing-coverts. Sometimes confused with Zenaida Dove. The endemic Plain Pigeon was once abundant and widespread in the Greater Antilles, but it is now an endangered species. Common on Hispaniola, but rare and local on Cuba and Jamaica, and recovering on Puerto Rico over past twenty years. It occupies a great variety of habitats from desert scrub to rainforest. Feeds in trees on fruits, seeds, buds and other vegetation. After the breeding season, gathers into flocks.

RING-TAILED PIGEON *Columba caribaea* 38–48cm

Large and pale grey with a black band across the middle of the tail; the tail is paler than the body, and the hindneck metallic green; no white on the wing, unlike Plain Pigeon. The legs are red, bill black and throat white. The only pigeon of the West Indies with a bicoloured tail. Endemic to Jamaica, and classified as critically endangered, albeit locally common. Inhabits mixed and coniferous mountain forest, feeding in flocks high in the canopy; can feed while hanging upside down. Eats only fresh fruit.

EURASIAN COLLARED DOVE *Streptopelia decaocto* 28–32cm

Individuals were introduced into the Bahamas in the early 1970s, and since this time the species has spread to Cuba and to islands of the Lesser Antilles. It is now becoming very common within its introduced range. A medium-sized, buff-grey dove with dark primaries and a collar round the hindneck which is black edged with white. It feeds on the ground and is frequently encountered in urban areas, where it finds a ready food supply.

WHITE-WINGED DOVE *Zenaida asiatica* 28–30cm

The White-winged Dove is a resident throughout the Greater Antilles and the Bahamas but is uncommon to rare elsewhere in the West Indies. It is undergoing some range expansion eastward within the area. Arboreal, and usually seen in flocks. Easily identified by the large white wing-patches which can be seen at rest as well as in flight, and by the white tips to all but the central tail-feathers. It has bare blue skin around the eye, though this is visible only at close range. Feeds mainly on seeds.

ZENAIDA DOVE *Zenaida aurita* 25–30cm

This chunky brown dove is a common permanent resident in the West Indies. It has white tips on the outer tail feathers; its underparts are variable in colour. In flight there is a conspicuous white trailing edge to the secondaries; the tail is rounded. A species of the open country, especially lowlands, but can be observed in woodlands. Feeds on seeds from the ground, but sometimes takes fruit from trees.

EARED DOVE *Zenaida auriculata* 22–25cm

The Eared Dove, also called the Violet-eared Dove, has colonised the West Indies from South America, and is a common species from St Lucia southward to Grenada in the Lesser Antilles. A short-tailed Mourning Dove lookalike, it is greyish-brown above and brown below. No white is evident on the wings or tail. It lives in arid scrub and avoids forested areas, though it perches in trees. Gathers into flocks, and feeds on seeds of grass and cultivated crops which it takes from the ground.

MOURNING DOVE *Zenaida macroura* 28–33cm

This brown dove, with upperparts displaying a purplish sheen, is easily distinguished by its long wedge-shaped and white-edged tail, but there is no white present in the wings. A permanent resident in the Bahamas and Greater Antilles, it is usually found in flocks in lowland, cultivated areas and open country. Increasing its range as a result of deforestation. In dry areas flocks will fly long distances to drink at dawn and dusk. Feeds on seed and other plant material taken from the ground.

COMMON GROUND-DOVE *Columbina passerina* 15–18cm

The smallest of the doves present in the West Indies, the male is grey above and scaly below, with a bluish-grey crown and hindneck; the female is a paler and rather smaller bird. Easily identified in flight by its small size, rapid, whirring flight and reddish wing-patches. Resident throughout the West Indies, it is a seed-eating ground-dweller of the forest, though it will sometimes take refuge in trees. Also inhabits open areas. It can be very tame around civilization, creeping mouse-like among the grass.

Male (above); female (below)

CARIBBEAN DOVE *Leptotila jamaicensis* 30–33 cm

The Caribbean Dove is locally common in Jamaica, New Providence in the Bahamas, and Grand Cayman. Walks rather than flies when disturbed. A chunky ground loving dove, with bright white face and underparts is most certainly this species. The legs are bright red; the underwings cinnamon. Its Caribbean range includes Yucatan and its islands, and islands off Honduras. Forages for seeds beneath shrubbery. Inhabits lowlands and foothills from forests to open areas, and is a permanent resident throughout its range.

KEY WEST QUAIL-DOVE *Geotrygon chrysia* 27–31cm

Rick and Nora Bowers

This stocky dove of the forest floor is generally secretive as it goes about its business. It has reddish-brown upperparts and whitish underparts. Head, nape, mantle and back have a green or purple iridescence, and there is a bright white stripe below the eye from the base of the bill to the nape. Inhabits understorey of dense forest, or dense thickets in arid areas, in the Bahamas, Cuba and Hispaniola; local in Puerto Rico. It is noisy as it forages for berries and other fruits in the leaf-litter, but takes to the wing silently.

54

BRIDLED QUAIL-DOVE *Geotrygon mystacea* 24–30cm

The Bridled Quail-dove is a multi-island endemic in the West Indies. It inhabits Puerto Rico, the Virgin Islands and the Lesser Antilles south to St Lucia. Duller and darker brown than Key West Quail-dove, it has brown upperparts with some iridescence. Easily distinguished in flight by the rufous patches in the wings. The upperparts as well as the underparts are brown. Feeds on the ground on seeds and snails. Lives in lowland forest, where it prefers drier areas than Ruddy Quail-dove.

GREY-HEADED QUAIL-DOVE *Geotrygon caniceps* 26–30cm

Endemic to Cuba and the Dominican Republic. This handsome light grey dove has a white forehead, purple back and a rufous vent and undertail-coverts; the bill and legs are reddish. It ranges throughout tropical lowland forests, and also invades coffee plantations in the Dominican Republic. Foraging, for seeds and invertebrates, takes place on the ground, and wet areas are favoured. This species is the only quail-dove within its range that has no face-stripe and is not brown. It has a swift and direct flight.

CRESTED QUAIL-DOVE *Geotrygon versicolor* 31cm

Rick and Nora Bowers

Endemic to Jamaica, this species is grey and rufous above and below; the elongated head feathers form a short distinct crest. There is a cinnamon stripe below the eye, a red eye-ring and bright red legs. Although seldom seen, it is a fairly common bird in some areas. Frequents undergrowth of wet limestone and montane forest, but is sometimes observed along roadsides. While walking it pumps its head forward and back, and flicks its tail up and down. Feeds on the forest floor scratching through the leaf litter for fallen seeds, fruits and invertebrates.

RUDDY QUAIL-DOVE *Geotrygon montana* 21–28cm

This, the most widely distributed of the quail-doves, is common in the West Indies in the Greater Antilles and in the heavily forested Lesser Antilles. The overall reddish-brown colouration with a buffy-white stripe below the eye is diagnostic. It is a ground-feeder, living in all types of forest at all elevations. It forages alone or in pairs, and feeds on seeds, fruits and small invertebrates. Most likely to be seen when it is flushed from forest trails early in the morning.

BLUE-HEADED QUAIL-DOVE
Starnoenas cyanocephala 29–34cm

Endemic to Cuba, and considered the most beautiful of the quail-doves, the endangered Blue-headed Quail-dove is also the largest of its group. A light blue head along with a white face-stripe, and black throat-patch are diagnostic; it is otherwise dark brown. Preferred habitat is thick-canopied forest with rocky soil and abundant leaf litter. The species is terrestrial in habits, and usually travels in pairs. Feeds on seeds, berries and snails. Flushes quickly when alarmed; the wings emit a quail-like whirring.

HISPANIOLAN PARAKEET *Aratinga chloroptera* 30–33cm

Endemic to Hispaniola, where it inhabits mountain woodlands. The bright colour of this large parakeet is not long seen, as it dives into trees and becomes invisible. There is a bit of red at the bend of the wing. Unlike the Hispaniolan Parrot, it has a long pointed tail and flies in a direct flight to its destination, usually in flocks. Known to place its eggs in abandoned termite nests. Feeds on seeds, fruits and farm crops, which causes farmers to kill it as a pest. Introduced into Puerto Rico and Guadeloupe, where it is local.

57

CUBAN PARAKEET *Aratinga euops* 24–27cm

Endemic to Cuba, this parakeet is smaller than the Hispaniolan and has less red on the wing bend. The long tail and direct flight are diagnostic for a parrot on Cuba. At rest one can observe widely separated specks of red in the feathers of the head and neck. This species is noisy in flight, but silent while feeding, taking a wide variety of seeds and fruits. It inhabits forests of all types as well as wet savannas. It is becoming rare in eastern Cuba.

ROSE-THROATED PARROT *Amazona leucocephala* 28–33 cm

The Rose-throated or Cuban Parrot is endemic to the Bahamas, Cuba and the Cayman Islands, and is the only West Indian amazon which isn't solely confined to one island. It is mainly green with a rose-red throat, foreneck and cheeks; the front of the head is white, primaries are blue and the belly maroon. Common in Cuba and Grand Cayman, less so in the Bahamas. In Abaco it nests underground in limestone crevices. A forest species at all elevations, it is threatened by habitat destruction and by the pet trade.

YELLOW-BILLED PARROT *Amazona collaria* 28–31cm

Endemic to Jamaica where it prefers montane evergreen forests. Like Jamaica's Black-billed Parrot it is predominantly green, but has a prominent yellow bill and maroon throat; it is more common and widespread than Black-billed. Being in high demand for the pet trade, it is classified as a threatened species. Its gentle temperament and brighter colours make it more popular than the Black-billed. Usually seen in pairs and small flocks sometimes along with Black-billed Parrots. Like other 'amazon' parrots, it is stocky, has a short rounded tail, and flies weakly with short, laboured wing-beats.

HISPANIOLAN PARROT *Amazona ventralis* 28–31cm

Easily distinguished by its bright green colour, white forehead, maroon belly and dark ear-spot. It is also the only species of amazon on Hispaniola, where it is endemic – though it has been introduced to Puerto Rico. As with the other Caribbean amazons, it is threatened by habitat destruction, as well as by hunting for food or for the pet trade, and is sometimes killed as a crop pest, though as yet it is relatively abundant throughout its range. The female of this species has been selected to foster the captive-laid eggs of the Puerto Rican Parrot. Feeds upon fruits and seeds.

PUERTO RICAN PARROT *Amazona vittata* 30cm

A green amazon with a white eye-ring, red forehead and two-toned blue primaries. One of the world's rarest birds, endemic to Puerto Rico. Can still be seen with luck in the Luquillo National Forest. Many factors have brought this parrot close to extinction, the worst being Hurricane Hugo which destroyed the forest in 1989. Collecting for the pet trade, predation by Red-tailed Hawk, and egg destruction by Pearly-eyed Thrasher are other causes. Recently the wild population has almost reached pre-Hugo levels, but Hurricane Mitch has gone through Puerto Rico recently and the current status is unknown.

BLACK-BILLED PARROT *Amazona agilis* 26cm

The rarest of the two endemic amazon parrots in Jamaica. It is entirely green with a blackish bill and white eye-ring. Occurring in flocks and in pairs, it probably feeds on different fare from the Yellow-billed Parrot since the two species occur in the same general areas. This species also occurs in tropical lowland forest. Fairly common in undisturbed areas like Windsor Caves, uncommon in eastern Jamaica. Unlike the Yellow-billed, it is not in demand for the pet trade since it is plainly coloured, and does poorly in captivity.

RED-NECKED PARROT *Amazona arausiaca* 33–36cm

Generally bright green in colour with a bright red foreneck. Known as the 'Jacquot' by local people, it is also endemic to Dominica. Occurs at lower elevations than Imperial Parrot, and as a result has suffered less, since more forest is available there. There are about 300 of these parrots left in the wild. The clearing of forest land for bananas has been, and is still, a severe problem for the species. It may be seen easily along the gorge at the Syndicate nature trail, sometimes along with Imperial Parrots.

ST LUCIA PARROT *Amazona versicolor* 42–46cm

The only parrot on St Lucia. A large yellowish-olive amazon, bright red on the throat and upper breast, with a blue head and yellow tail; bright red wing-patches are prominent in flight. Endemic to central mountain forests of St Lucia, where its population has grown from less than 100 birds in 1977 to over 350 today. Relatively common at the Union Nature Center where there is widespread mountain rainforest. Flocks of up to 20 birds feed on fruit in the forest canopy, and roost communally. The national bird of St Lucia and a beneficiary of the public education programme.

ST VINCENT PARROT *Amazona guildingii* 41–46cm

The only parrot on St Vincent. Its golden-brown to green plumage and yellow-tipped tail with a broad blue subterminal band make this endemic parrot one of the world's most beautiful. Fairly common in the Vermont Nature Center. A successful captive-breeding programme is being carried on at the Nicholl's Wildlife Complex within the botanical garden. This species prefers mature moist forest to 1,000m, but will descend to lower elevations for nesting. Flocks and roosts communally. Feeds on fruits, seeds and blossoms in the treetops. The national bird of St Vincent.

IMPERIAL PARROT *Amazona imperialis* 46–51cm

The 'Siserou' as it is known locally is the largest amazon, and second rarest of West Indian parrots. It is endemic to Dominica, and is the national bird. In the past, the population has dropped due to hurricanes, trapping and hunting, and habitat loss now continues the decline. Only about 50 birds remain in the wild. It is large and purple, looking black in flight, and occurs in mixed flocks with Red-necked Parrots. Can be seen along the gorge at the Syndicate nature trail. Normally occurs above 450m, in the canopy of primary rainforest of mountain slopes.

YELLOW-BILLED CUCKOO *Coccyzus americanus* 28–32cm

Francisco Rivas

An uncommon breeder in the Greater Antilles and Bahamas, and a common migrant elsewhere in the West Indies. Like the Mangrove Cuckoo, it has a long white-tipped tail and a downcurved bill with a yellow lower mandible, but, unlike it, it has white underparts, no black on the cheek, and a reddish-brown wing-patch which shows in flight. In the West Indies it frequents scrub and dry forest in the lowlands. An inactive species, moving slowly as it searches for insects, especially caterpillars, and it sometimes feeds on frogs and lizards.

MANGROVE CUCKOO *Coccyzus minor* 28–30cm

A fairly common resident throughout the West Indies and similar in general appearance to Yellow-billed Cuckoo, being slender-bodied with a white-tipped tail and a long decurved yellow-based bill – but Mangrove Cuckoo's black ear-patch and buffy abdomen are diagnostic. Its habitat varies from mangroves to dry scrub, but its range does not extend into the high mountains. This rather sluggish species is usually located by its low, guttural 'gawk gawk gawk' calls. Feeds on insects, especially caterpillars.

GREAT (CUBAN) LIZARD-CUCKOO
Saurothera merlini 44–55CM

This huge cuckoo is endemic to Andros, New Providence and Eleuthera in the Bahamas, as well as to Cuba and its islands. Ungainly, sluggish and tame, it flourishes in thick undergrowth. The Cuban form has dark reddish-brown wings, whereas Bahama birds lack the reddish. The upperparts are brownish-grey, underparts whitish and the belly rufous to buff. There is a patch of bare red skin round the eye, and a long, straight grey to brown bill. Solitary and tame in Cuba; secretive in the Bahamas. Forages in mid-levels of vegetation for lizards, small snakes, fruit and mice.

JAMAICAN LIZARD-CUCKOO *Saurothera vetula* 38–40cm

This is the smallest of the lizard-cuckoos. Endemic to Jamaica where it forages in lowlands, and in brushy and forested areas at altitudes of up to 1,200 m. It feeds on lizards, insects and even on nestling birds, and has been observed probing into nesting holes and crevices for its prey. Though similar to Chestnut-bellied Cuckoo, another endemic resident of Jamaica, it can be distinguished by its smaller size, long straight bill, pale grey breast, rufous patch on the wings and bare red skin around the eye.

HISPANIOLAN LIZARD-CUCKOO
Saurothera longirostris 41–46cm

The most colourful of the lizard-cuckoos, known to the local people as 'Pajaro bobo' or crazy bird, the Hispaniolan Lizard-cuckoo is endemic to Hispaniola where it is widespread in the Dominican Republic and less common in Haiti. A large bird, grey above, with chestnut wing-patches, pale grey breast, long white-tipped tail, long bill and bare red skin around the eye. Occurs from sea-level to 2,000 m. Forages throughout the understorey, and into the canopy to feed on insects, lizards and small snakes. The flesh of this species is locally said to promote appetite and aid digestion.

PUERTO RICAN LIZARD-CUCKOO *Saurothera vieilloti* 40–48cm

This lizard-cuckoo is known in Puerto Rico as 'Pajaro de aqua' (water bird) because its call is believed to forecast rain. An endemic species which is a common though secretive resident throughout most forested areas of the country. Its large size, long tail, two-tone underparts (grey and cinnamon) and bare red skin around the eye permit easy identification. Feeds on lizards and insects and spiders, sometimes prying bark from trees in its search.

CHESTNUT-BELLIED CUCKOO *Hyetornis pluvialis* 48–56cm

This endemic Jamaican cuckoo is very large and grey with chestnut underparts, a whitish throat, grey breast and long grey tail-feathers terminating in large white spots. The heavy dark grey bill is downcurved, and is much shorter than that of Jamaican Lizard-cuckoo. Runs along branches gliding from tree to tree. Feeds on lizards, mice, insects, eggs and the nestlings of other birds. Frequents open wet forest at mid-elevations, and woodland near cultivated areas in the hills and mountains.

SMOOTH-BILLED ANI *Crotophaga ani* 30–33cm

The glossy black plumage with a heavy deep bill and long, flat tail are distinctive features. Occurs in small noisy flocks, which tend to fly in lines with weak, flappy flight. The group roosts by squeezing together on a perch in order to remain in contact. Native to the Bahamas, Greater Antilles, Virgins and Caymans. It is rare to common in the Lesser Antilles. Nests are built and used communally by a number of females, and the young are cared for and fed by the group as a whole. Feeds on insects, lizards and frogs.

BARN OWL *Tyto alba* 30–43cm

This most cosmopolitan of the owls is common throughout the West Indies. Its pure white underparts (in some races), silent flight and piercing scream have given rise to haunted house legends. The white, heart-shaped face and pale brown upperparts add to this frightening apparition. Nocturnal, it feeds mainly on rodents, but has been known to take petrels as they return to their nests. Perches on fence posts, especially around rice fields, dry scrubland and open woodland. Usually seen at dusk gliding over open land.

ASHY-FACED OWL *Tyto glaucops* 35cm

The Ashy-faced Owl, endemic to Hispaniola, is similar to, but smaller than, the Barn Owl. With the exception of that very widespread species, it is the only other member of this family (Tytonidae) in the western hemisphere. It lives in open, coastal lowland forests. Locally common, it is nocturnal and feeds upon bats, rodents, lizards, frogs and birds. Its heart-shaped face is ashy-grey rather than pure white as in Barn Owl. Roosts in limestone cliffs.

PUERTO RICAN SCREECH-OWL *Otus nudipes* 23–25 cm

Endemic to Puerto Rico where common, and the Virgin Islands where it is rare. Greyish-brown above, marked with heavy brown streaks below, ear tufts unnoticeable unless the bird is alarmed. Nocturnal, lives in dense forests, coffee plantations, and isolated dense thickets on the coast. Feeds on insects. Entirely nocturnal, but very responsive to a recording of its voice, which is a screech-owl type trill. Natives believe that this species eats coffee beans, and that eating its flesh can cure asthma. Roosts in caves or dense vegetation by day. Grey phase rare.

R. A. Behrstock

CUBAN SCREECH-OWL/BARE-LEGGED OWL
Otus (Gymnoglaux) lawrencii 20–23cm

R. A. Behrstock

Formerly called Bare-legged Owl, it perches upright like a Burrowing Owl, and resembles it, but the underparts are streaked, not barred. Like Burrowing Owl it has no ear-tufts. The eyes are dark and the legs bare. Feeds on large insects, frogs and small birds. Endemic to Cuba, in tropical lowland and mountain evergreen forest, where it roosts and nests in holes in palms, and on limestone cliffs. Scratching gently on the trunk of a roost-tree will usually entice the owl to stick its head out of the hole.

CUBAN PYGMY OWL *Glaucidium siju* 17.5cm

This diurnal owl is endemic to Cuba. It cocks its tail upon perching, and sometimes swings it from side to side. The smallest raptor in Cuba, this tiny owl is widespread, common and very tame. Its short yellow feet are covered with feathers; the tail is short. Two dark spots are evident on the rear of the head resembling eyes. The upperparts are grey to brown; the underparts are white streaked with tan. Its preferred habitat varies from deep woods to open countryside and banana plantations. It may be seen hunting at any time of the day.

STYGIAN OWL *Asio stygius* 41–46cm

A large dark brown owl with noticeable ear tufts and yellow eyes. The Stygian Owl is fairly common in Cuba, but is an extremely rare bird in Hispaniola, and is endangered by deforestation throughout its range. Usually found in mountain forests, but easily adapts to other habitats. It does not begin to hunt until well after nightfall, and feeds mainly on bats and on roosting doves. In Cuba, the Stygian Owl is considered an evil omen and is killed for that reason.

SHORT-EARED OWL *Asio flammeus* 35–43cm

A medium-sized owl, brown above and paler below with heavy streaking on the breast which fades on the abdomen; no ear-tufts present. In flight it displays black wrist-patches on white underwings and buffy patches on the brownish upperwing. It flies moth-like, low over fields, sometimes perching on fenceposts, but roosting on the ground. Mainly diurnal, though it is most active at dusk. Common on Cuba and Hispaniola, less so on Puerto Rico. Feeds on small mammals, and hunts pastures, marshes, rice fields and other open areas.

JAMAICAN OWL *Pseudoscops grammicus* 31–36cm

Joel Greenberg

This yellowish-brown owl is endemic to Jamaica. It is a medium-sized species with ear-tufts that are conspicuous. Widespread at all elevations, and in all habitats, it occurs anywhere that there are trees suitable for it to nest in. Uses the same daytime perch regularly. Feeds on mice, insects, lizards and tree-frogs. The Jamaican Owl is strictly a nocturnal species. Its growling 'waugh' call is distinctive, and can be heard at dusk and before dawn.

ANTILLEAN NIGHTHAWK *Chordeiles gundlachii* 20–25cm

Similar to Common Nighthawk *C. minor* which is a migrant throughout the West Indies, whereas the Antillean is a breeding resident only in the Bahamas, Cayman Islands and Greater Antilles. It is difficult to separate the two species in the field, but Antillean Nighthawk has tan (rather than black) underwing-coverts and its flight is more erratic; its call in flight is 'pity pat pit' while Common Nighthawk utters a nasal 'peent'. Sometimes occurs in large flocks on overcast days. Roosts on the ground, or lengthwise on a horizontal branch. Dives swiftly for moths and beetles.

CUBAN NIGHTJAR *Caprimulgus cubanensis* 28cm

Endemic to Cuba, it lives in dense wet forest and scrub, and prefers wood and field edges and roads. A large dark nightjar (sexes are alike), it is crepuscular and nocturnal. The upperparts are greyish to blackish-brown; there is no collar on the hindneck, and no white markings on the wings. The introduced Small Indian Mongoose is a serious predator. It was formerly known as Greater Antillean Nightjar, but has recently been separated from the Hispaniolan form.

PUERTO RICAN NIGHTJAR *Caprimulgus noctitherus* 22cm

R. A. Behrstock

An endemic species thought to be extinct since 1888, but rediscovered in the dry forest of south-west Puerto Rico in 1961. It is nocturnal, and difficult to see well, since it usually circles once or twice then disappears into the brush. It is a small greyish-brown goatsucker; the male has large white patches on the outer tail. The 'weep weep' call is reminiscent of a guinea pig. Sallies into forest openings and under the canopy to feed on nocturnal insects. The migrant Chuck-will's-widow *C. carolinensis* is much larger; Antillean Nighthawk has white bands on the wing.

NORTHERN POTOO *Nyctibius jamaicensis* 43–46cm

A large-headed, long-tailed night bird, dark brown and cinnamon streaked with buff. The large yellow eyes reflect brilliant red when exposed to a bright light source. Its rudimentary bill belies the enormous size of its mouth when opened. Perches upright and immobile on a broken stub or fence post when roosting or when alarmed, but sallies forth to feed like a huge flycatcher. Large beetles and moths are the normal fare. It is common on Jamaica, but rare on Hispaniola. Its habitat is forests near open areas, but golf courses and palm groves are also frequented. Its call is a loud guttural growl.

ANTILLEAN MANGO *Anthracothorax dominicus* 11–12.5cm

A large hummingbird, yellowish-green above. The male has blackish underparts and a brilliant green gorget. The female is the only hummingbird in its range with whitish underparts, and white tips to the tail-feathers. Black bill slightly downcurved. Endemic to the West Indies: common on Hispaniola, frequents dry coastal areas of Puerto Rico, and is rare in the Virgin Islands. Prefers open areas and scrub; frequents gardens and coffee plantations. Feeds on nectar from a variety of flowers, but also hawks insects from a perch. Declining due to encroachment by the Green-throated Carib.

Male (above); female (below)

GREEN MANGO *Anthracothorax viridis* 11–14cm

The Green Mango is endemic to the central and western mountains of Puerto Rico; rare in the northeast, and in coastal areas. It is a large, principally green species, which is common in coffee plantations, dense forests, and edges of the interior; sometimes observed along the western coast of Puerto Rico. The underparts are entirely emerald green, the black beak is down-curved, and the tail is rounded; sexes are alike. Feeds upon animal matter and nectar. Hawks insects above the canopy, and gleans spiders from leaf surfaces. The male defends feeding sites around flowering plants.

JAMAICAN MANGO *Anthracothorax mango* 13cm

Endemic to Jamaica, this large dark hummingbird has a long black decurved bill. When observed in good light the black underparts together with the magenta on the cheeks and the sides of the neck are spectacular. It is abundant near the coast and in other lowland country, including banana plantations and gardens. Feeds on insects and the nectar of a great variety of flowers. This is the only Jamaican hummingbird which feeds extensively at cactus flowers, and pulls insects from spider webs.

PURPLE-THROATED CARIB *Eulampis jugularis* 11.5cm

This large hummingbird appears completely black at a distance, especially when not seen in full sunlight. The gorget is purplish-red, and the tail and uppertail-coverts are bluish-green. The green wings are diagnostic in the field. Females have longer bills than males, though the reason for this has not been ascertained. It is a bird of mountain forests and openings and will sometimes frequent banana plantations. A widespread species of the Lesser Antilles. Feeds on nectar and insects, which are hawked from the air.

GREEN-THROATED CARIB *Eulampis holosericeus* 10.5–12cm

Endemic to the West Indian islands of the Lesser Antilles, Puerto Rico and the Virgin Islands. It is a fairly large hummingbird, with a distinctive downcurved bill (which is longer in the female than in the male), a green breast and a violet-black tail. A blue breast mark can be seen under ideal lighting conditions. Generally a smaller and less robust bird than the Purple-throated Carib; the two species are often to be seen feeding close together. Frequents gardens and forests, and is most commonly seen at lower elevations.

ANTILLEAN CRESTED HUMMINGBIRD
Orthorhynchus cristatus 8.5–9.5cm

This small hummingbird is easily recognised by its crest which displays green and blue iridescence. It is green above and grey below, and has a short, straight bill. One of the most common birds of the Lesser Antilles, it is also present in north-east Puerto Rico and in the Virgin Islands, and is to be found in all habitats from sea-level up into the mountains. It is a species which prefers nectar, but the birds will feed on insects and spiders during the dry season.

EMERALD HUMMINGBIRDS *Chlorostilbon spp.* 9–11cm

Cuban Emerald (left); Puerto Rican Emerald male (centre), female (right)

The Cuban, Hispaniolan and Puerto Rican Emeralds are nearly identical hummingbirds, all with long, black, forked tails, and are endemic to their respective islands. Males are mainly bright green above and below: the Cuban has white undertail-coverts, the Hispaniolan is black-breasted and the Puerto Rican has completely green underparts. The various females have a forked tail, white to grey underparts with green on the flanks, a black bill and white-tipped outer tail-feathers. The species are widespread in all habitats, and feed on nectar and insects.

BLUE-HEADED HUMMINGBIRD *Cyanophaia bicolor* 9.5cm

The Blue-headed Hummingbird occurs on only two islands of the Lesser Antilles – Dominica and Martinique – usually above 600m. A species of elfin forest and rainforest understorey, it is fond of bathing along mountain streams. The male's head, back and wings are black; blue extends onto the throat and breast. The lower mandible is pink with a black tip. The female has a blackish ear-patch, but otherwise has green and blue upperparts, and is whitish below. Feeds on nectar and small insects near ground level.

BLACK-BILLED STREAMERTAIL *Trochilus scitulus* 22–24cm

Endemic to extreme eastern Jamaica, where the Red-billed Streamertail is absent. The Black-billed has a black crown and ear-tufts; the distinctive black bill is thinner at the base. The long tail-streamers of the male are distinctive. The female lacks the streamers, has white underparts and is only half the length of the male. Feeds on nectar and insects in humid forests, banana plantations and gardens; sometimes feeds on a flower after Bananaquits have pierced the corolla. Formerly combined with Red-billed Streamertail, from which it differs in its call, in the colour and width of its bill, and in the courtship flight.

RED-BILLED STREAMERTAIL *Trochilus polytmus* 22–25cm

Female (right); male (far right)

The most common bird in Jamaica, ranging from arid lowlands to the mountains, and the most spectacular hummingbird of the West Indies. The male has a red, black-tipped bill, and long tail-streamers, which due to their shape produce a hum in flight. When perched these streamers are usually crossed. The body is iridescent green, the wings brown and the tail black; a black crown and ear-tufts are evident. The female is green above and white below, with no streamers. Feeds on nectar and small insects. Endemic to Jamaica, being absent only from the extreme east, where it overlaps with Black-billed Streamertail.

BAHAMA WOODSTAR *Calliphlox evelynae* 9–9.5cm

Male

Female

This small, aggressive hummingbird is endemic to the Bahamas. The male is green on the back and wings, as well as on the head; its underparts are white and the throat magenta; the tail is forked. Females lack the magenta throat and have a rounded tail. The woodstar is rare on islands where the Cuban Emerald is abundant (Grand Bahama and Abaco), as it is bullied by it. It feeds mainly on nectar and insects in gardens, woodland edges and coppices. Very tame, it perches on twigs and wires.

VERVAIN HUMMINGBIRD *Mellisuga minima* 6cm

This species, the world's second-smallest bird, is endemic and common in Jamaica and Hispaniola. The head and back are green, and the white underparts are speckled with green along the sides of the breast and throat. The male has a distinct twittering song which it sings from an exposed perch. A bird will hover with tail cocked as it feeds on small flowers; red ones are preferred. It occasionally hawks insects from a perch. The favoured habitat is forest edges, gardens and roadsides. Very aggressive towards other bird species.

BEE HUMMINGBIRD *Mellisuga helenae* 5.5cm

This species is the smallest bird in the world; endemic to Cuba. The male has an iridescent red crown and throat which has side-plumes during the breeding season. The female (shown here) lacks these features. Upperparts are bluish, underparts greyish-white. Tame and solitary, tending to perch high in trees. Prefers horizontal flowers, especially majagua and aloe, but when flowers are in short supply it feeds on insects. Frequents forests and their edges, also swamps and gardens. The wing-beats resemble those of a bumble-bee.

CUBAN TROGON *Priotelus temnurus* 25–28cm

This spectacular species, the national bird of Cuba, is endemic to the island, and is widely distributed and common. The red belly, green back, blue crown and short broad bill are diagnostic. The tail, white below, is long and ragged-looking due to its peculiar-shaped feathers. Birds are usually found in pairs in shady areas of both wet and dry forests. Feeds on flowers, buds and fruits which it obtains by hovering after a short noisy flight. This trogon has a characteristic posture when perched; its call is distinctive.

HISPANIOLAN TROGON *Priotelus roseigaster* 27–30cm

Front   *Back*

Endemic to Hispaniola where it is the only trogon; fairly common in suitable habitat. Its numbers are declining, particularly in Haiti, due to habitat destruction. Diagnostic glossy green upperparts, red belly, yellow bill and grey throat and breast, along with long dark blue tail marked with white below. The male has fine black-and-white barring on the wings. Occurs in mountain forests, both pine and broadleaf, usually in pairs. Feeds mainly on insects and small lizards as well as on small fruits which it takes in flight. Nests in old woodpecker holes.

CUBAN TODY *Todus multicolor* 11cm

The Cuban Tody is the most beautiful of this group. The only tody in Cuba, and widely distributed in wooded areas, especially near earthen banks in which it excavates its nesting burrows. Usually occurs in pairs. Feeds on insects, and has a voracious appetite; eating 40% of its body weight daily. The bright green colouration of the upperparts along with the bright red throat make it a favourite of the local people. The sides of the neck are blue. The characteristic call 'tot-tot-tot-tot' makes it easy to locate in any habitat.

BROAD-BILLED TODY *Todus subulatus* 11–12cm

Adult   Immature

Two species of tody are endemic to Hispaniola. The Broad-billed Tody lives in arid areas and secondary growth, and is distinguished from the similar Narrow-billed Tody by its wholly red lower mandible and slightly larger size; its green upperparts are washed with a yellowish-gold cast. In habits it is less active, and forages higher. The distinctive call is 'terp terp terp'. Todies perch with the head elevated, searching the bottom of leaves for insects, especially caterpillars, which they snatch on the wing.

NARROW-BILLED TODY *Todus angustirostris* 11cm

The Narrow-billed Tody is also endemic to Hispaniola. It prefers mature upland forest and coffee plantations. When seen closely, the bill is longer and more slender than Broad-billed Tody's, while the tip of its lower mandible is black. The upperparts are bright dark green. The two-syllable call, 'chip chee', is very unlike Broad-billed Tody. A bird will change its lookout perch frequently. It sallies to catch its prey in flight, and thrashes it against a branch before swallowing it. Todies relish bees, and can become a pest to beekeepers.

JAMAICAN TODY *Todus todus* 9cm

Endemic to Jamaica, and the only resident tody there, it is the smallest of the family. Widespread and common in all habitats from the coast to the mountains, it is known as 'robin redbreast' by the local people. Todies are bright green, and adults have a bright red throat. A tody can be attracted by finger-snapping, or clicking two rocks together which simulates bill-snapping. In flight the tody makes a rattling sound which is caused by the narrow-tipped primaries.

PUERTO RICAN TODY *Todus mexicanus* 11cm

Endemic to Puerto Rico where it is the only species of tody present. Its call is a nasal 'beep'. One to 4 white eggs are deposited in the underground nesting chamber. Todies are related to motmots and kingfishers, and the family is one of only two endemic to the West Indies. They have some of the highest rates of feeding young ever recorded for insectivorous species. Young todies lack the red throat of the adult, and have dusky green streaks on the breast, and brown eyes.

RINGED KINGFISHER *Ceryle torquata* 38–41cm

This is the largest kingfisher of the Americas, and in the West Indies is a resident in Guadeloupe, Dominica and Martinique. It has a crest and a wide white collar. The upperparts are slate-grey, and the underparts reddish-brown, entirely so in the female; the male has a white abdomen. The tail is banded black and white; in flight, reddish underwing-coverts show. It is similar to Belted Kingfisher *Ceryle alcyon*, but larger and with a massive bill. Dives for fish along the edges of fresh water, and along the coast where streams flows into the sea.

GUADELOUPE WOODPECKER *Melanerpes herminieri* 24–29cm

The only breeding woodpecker found on Guadeloupe in the Lesser Antilles, where it is both endemic and common. Though appearing completely black at a distance, it has a reddish tint when seen at close range. The bill of the male is considerably longer than that of the female. Shy, it quietly feeds on insect larvae which it searches for on dead trees. Also feeds on fruit and tree-frogs. It lives in all the habitats present, but is more common on Basse-Terre. Doesn't stay long at feeding sites, but regularly visits fruiting trees.

Phillipe Feldmann

PUERTO RICAN WOODPECKER
Melanerpes portoricensis 22–27cm

This very colourful woodpecker has dark upperparts, a white rump, chestnut undertail-coverts, a red breast and throat and a white forecrown which encompasses the eye. It is endemic to Puerto Rico and occurs in almost all habitats, most commonly in mountain coffee plantations. Its principal food is wood-boring beetles, ants and earwigs, but seeds and fruit are also important. Males peck and probe as they forage in lower and middle sections of trees; the female feeds in mid-levels and in the canopy. Commonly nests in telegraph poles.

HISPANIOLAN WOODPECKER *Melanerpes stratus* 20–25cm

A woodpecker with a red crown and rump, grey face and brownish-olive underparts with white stripes on the hindneck; the back is barred yellow and black. The endemic Hispaniolan Woodpecker is the most widely distributed species on the island. The male has a noticeably longer bill than the female. Forms noisy social groups, and sometimes nests in colonies. Feeds on plant and animal life in equal amounts, by probing, pecking or gleaning. Can hang upside down when feeding, and sometimes hawks insects from a perch.

JAMAICAN WOODPECKER *Melanerpes radiolatus* 24–26cm

The Jamaican Woodpecker has a white face, a red crown and a reddish-brown breast, and its black upperparts are finely barred with white. It is endemic to Jamaica but is widespread there, being a common bird at all elevations. It feeds on the epiphytic plants which trees support, as well as on the trees themselves. The diet is chiefly surface-living insects, and wood-boring species are rarely taken. Fig fruits are another important food.

WEST INDIAN WOODPECKER
Melanerpes superciliaris 26–32cm

This most common Cuban woodpecker is a West Indies endemic which also occurs on some of the Bahamas (Abaco and San Salvador) and on Grand Cayman; it is extremely rare on Grand Bahama. The upperparts are barred black and white; the male has a red crown and hindneck, the female is red only on nape. The abdomen is red. Frequents wooded areas, dry forest, scrub, coppice and palm groves. Gleans arthropods from bromeliads; also feeds on the ground. It is frugivorous to a great degree.

CUBAN GREEN WOODPECKER
Xiphidiopicus percussus 21–25cm

The most abundant of the Cuban woodpeckers, it is also an endemic. It has a large crest, the upperparts are green, the underparts are yellow and there is a red spot on the breast. The face is white with a black stripe through the eye. Feeds on very large insects. Lives in forest and woodland at any elevation, also in mangroves and palms. Searches thick vines for insects, or feeds on dead branches, where it preys on insects taken from beneath the bark.

HAIRY WOODPECKER *Picoides villosus* 20–23 cm

Male   Female

This North and Central American species in the West Indies occurs only in the northern Bahamas, where it is a permanent resident. An inhabitant of the pine forests; it can be heard drumming loudly on trees. Feeds on the lower trunks of large trees, but is sometimes observed on the ground. When foraging in pairs, males feed on cones, females on trunks. Pine seeds and insects are staples. The Hairy Woodpecker is black above and white below; the male has a red spot on the back of the head which is absent in the female. The moustache stripe as well as the eyestripe is black.

NORTHERN FLICKER *Colaptes auratus* 30–32cm

This species is resident in Cuba and the Cayman Islands, and is being considered for new species status. Frequents open forest and edges, and anywhere near open ground. Feeds commonly on the ground; it probes for ants, which are its primary food source, but also feeds on fruits, berries and nuts. In flight it displays bright yellow underwings, and a white rump spotted with black. When perched it is a large woodpecker with a black bar across the breast, pale buff underparts spotted with black, and a red patch on the hindneck. The male has a black moustachial mark.

FERNANDINA'S FLICKER *Colaptes fernandinae* 33–35cm

This second-largest Cuban woodpecker, endemic to the island, is yellowish-tan with fine black barring; the underwings are yellow. Unlike the Northern Flicker there is no red on the head. Fernandina's Flicker is a threatened species due to loss of habitat, and is now common only near the Zapata Swamp, though it was once widespread. A ground-feeder of savanna, pasture, swamp and various types of woodland, especially areas near palms, which are preferred as nest-sites. Feeds on insects, grubs, earthworms and seeds which are taken from the leaf litter by probing.

CARIBBEAN ELAENIA *Elaenia martinica* 15.5–18cm

The Caribbean Elaenia is common and widespread throughout Puerto Rico, the Lesser Antilles, Caymans, some of the Bahamas and the Netherlands Antilles. It is olive-grey above, has two whitish wing-bars and the greyish-yellow underparts are washed with yellow. There is a suggestion of a crest which, when erected, is whitish-yellow. The lower mandible is pink. It occupies woodland and scrub areas where it prefers dry lowland, but ranges into the mountains. Calls well into the day.

YELLOW-BELLIED ELAENIA *Elaenia flavogaster* 17cm

Despite the bird's name, its underparts are only slightly yellow. It is brownish-yellow above, with a conspicuous bushy crest which it tends to raise slightly, sometimes displaying a white patch in the centre of the crown. There are two whitish wing-bars which have yellow-white edging, and a white eye-ring. Occurs in the West Indies only on St Vincent, the Grenadines and Grenada. Frequents open country in the lowlands where it feeds on insects and small berries. A noisy species which forages in pairs.

CRESCENT-EYED PEWEE *Contopus caribaeus* 15–16.5cm

This species was formerly called Greater Antillean Pewee and is endemic to the island of Cuba and the Bahamas. The upperparts are dark brownish-olive and the underparts greyish-buff. There is a distinct white crescent behind the eye; there are no wing-bars. It trembles its tail upon landing on a perch. Frequents pine forest, edges of clearings, mangroves and orchards. Feeds on fruit and on insects caught in flight by sallying forth from a favourite perch.

JAMAICAN PEWEE *Contopus pallidus* 15cm

The endemic Jamaican Pewee has dark olive-brown upperparts, a darker head and buffy-brown underparts. It can be distinguished from the other flycatchers within its range by its lack of distinct wing-bars. Its lower mandible is pale orange, and its tail, which is flicked upon landing, is notched. Catches insects in the air by sallying from an exposed perch; it travels to different perches within its territory. Others Jamaican flycatchers take insects from leaf surfaces. Common and widespread in forest edges, and mid- to high-elevation forests.

HISPANIOLAN PEWEE *Contopus hispaniolensis* 15–16cm

Adult   Immature

Endemic to the island, the Hispaniolan Pewee has grey-olive upperparts, which are darker on the head. The underparts are grey with an olive to brown wash. Indistinct wing-bars and a pale lower mandible are diagnostic. Flicks its tail upon landing. Common from the coast to the mountains, it catches insects on the wing. It is tame, and returns to a familiar perch, usually in the understorey; also feeds on fruits. Widespread within different habitats including forests and their edges, coffee plantations and orchards. In pine forest it joins mixed flocks.

ST LUCIA PEWEE *Contopus oberi* 15cm

This small endemic pewee of St Lucia has dark olive-brown upperparts. The underparts, including the throat, are reddish-brown, and the bill has a yellow lower mandible. Like the similar forms of Hispaniola, Jamaica and Puerto Rico, this bird was recently split off as a species separate from the Crescent-eyed Pewee. It occurs commonly in forest clearings at higher altitudes, but is rare elsewhere. Sallies from a low perch, and snatches insects on the wing.

LESSER ANTILLEAN PEWEE *Contopus latirostris* 15cm

Adult

Immature

This species is endemic to Guadeloupe, Dominica and Martinique, and possibly St Kitts and Nevis, and is yet another split from the Crescent-eyed Pewee. Within its range it is a common but local species. The upperparts are brownish-olive, the wings and tail are black and the underparts yellowish-brown. The lower mandible of the bill has a pale base. The species chiefly frequents mountain forests and woodlands, and is uncommon elsewhere. It chooses a favourite perch at a low level from where it sallies out to catch insects on the wing.

SAD FLYCATCHER *Myiarchus barbirostris* 16cm

Endemic to Jamaica, and an abundant resident, the Sad Flycatcher is the smallest of the three *Myiarchus* species present in Jamaica. The upperparts are brownish-olive, the crown is dark, the throat is white and the remainder of the underparts are yellow. There is buff edging to the wings and their coverts which can resemble wing-bars. Frequents forests and wooded areas of the lowlands up to mid-altitude. Sallies from a favourite perch to collect insects from the foliage, sometimes taking several before returning to its perch.

GRENADA FLYCATCHER *Myiarchus nugator* 20cm

Endemic to St Vincent, Grenada and the Grenadines, this flycatcher has olive-brown upperparts, a dark brown head, a black bill with a pink lower mandible, and two pale brown wing-bars. The primaries and the outer tail-feathers have reddish edges. The secondaries are fringed with grey or cream, and the inside of the mouth is bright orange. A permanent resident at all elevations, but prefers open to lightly wooded areas, and lowland scrub near palms. Common around villages. Sallies from its perch to collect insects, flicking its tail upon landing. Also feeds on fruit.

LA SAGRA'S FLYCATCHER *Myiarchus sagrae* 19–22cm

This is a medium-sized *Myiarchus*, endemic to the Bahamas, Cuba, and the Caymans. A common species except in the southern Bahamas. Its manner of perching is unique, as it leans forward at a 45° angle. The flat-headed appearance, black bill and faint wing-bars are diagnostic. Upperparts are greyish-olive; the rump is cinnamon during breeding, and grey otherwise. The crown is dark; sides of the neck and breast are grey, and the underparts greyish-white. It raises its crest when disturbed. Frequents understorey of pine and mixed woodlands, at all elevations; sometimes lives in mangroves.

STOLID FLYCATCHER *Myiarchus stolidus* 20cm

This common species is endemic to Jamaica and Hispaniola, and their islands. A medium-sized *Myiarchus* flycatcher, with two white wing-bars. The head is olive-brown, and the rest of the upperparts grey. The wings are brown with brighter markings than in Sad Flycatcher. Its habitat is decidedly different from the other *Myiarchus* of Jamaica as it lives in arid woodland, mangroves and scrub, as well as lowland forests and their edges. An active and tame species, it feeds on insects and berries, taken while in flight.

PUERTO RICAN FLYCATCHER *Myiarchus antillarum* 18–20cm

Endemic to Puerto Rico and its islands, and to some of the Virgin Islands. Formerly considered the same species as Stolid Flycatcher, from which it has been separated because of its call. The upperparts and head are dark brown; the underparts are brownish-grey fading to white on the undertail, and there is no yellow wash. The wing-bars are indistinct. Its population was severely damaged by a 1930s hurricane but it is now common. Frequents wooded areas, borders of mangroves, scrub, coffee plantations and mid-level mountain forest, where it feeds on insects and small fruits.

GREY KINGBIRD *Tyrannus dominicensis* 22–25cm

Grey above and white below, and has a black mask which extends from its large bill under the eye to the ear-coverts. The tail is notched. An abundant species in the West Indies, commonly seen on wires and on the tops of scattered trees. Feeds on large insects, which it catches on the wing; it then returns to its perch, and pounds its prey on a hard surface to soften it. It sometimes feeds at night under street lights. This kingbird is a very aggressive species, and sometimes forms large communal roosts.

LOGGERHEAD KINGBIRD *Tyrannus caudifasciatus* 24–26cm

Endemic to the northern Bahamas, Greater Antilles and Caymans, but rare on Little Cayman. It has dark upperparts with a darker crown. The black of the crown extends below the eye; the underparts are pure white. The square tail is black and tipped with off-white, except for birds in Puerto Rico and Hispaniola. This flycatcher has a very large black bill. Frequents more forested areas than Grey Kingbird, including pine and broadleaf woodland, coffee plantations and mangroves. It sallies for insects and small lizards from low exposed perches, or often from telephone lines.

GIANT KINGBIRD *Tyrannus cubensis* 23cm

This rare Cuban endemic was formerly a resident in the Bahamas too, but it is now extinct there. It has a very large bill, the upperparts and crown are very dark and the underparts are white; the tail is notched. The similar Grey Kingbird occupies similar habitat. Ceiba trees seem to be important to this species. The kingbird roosts high in these giants, where it sallies after insects, in a sluggish manner. Frequents forest and woodlands near swampy areas; also semi-open areas and pineland. Feeds also on lizards, fledglings and fruit.

JAMAICAN BECARD *Pachyramphus niger* 18cm

Endemic to Jamaica, and the only becard in the West Indies; it is locally common. The male is glossy black above, and lighter below. Its head is large, the bill thick, the tail short, and it displays a white mark at the wing-base in flight. The female is rufous-brown above and pale buffy-grey below with cinnamon cheeks and throat. It travels slowly through the canopy as it feeds on insects and fruit, and also forages into the mid-levels. Sometimes hawks insects from a perch. Frequents edges and openings of tall forests at mid elevations, also pastures and gardens.

THICK-BILLED VIREO *Vireo crassirostris* 14cm

This vireo is a resident of the Bahamas, and is local in Cuba, Hispaniola, the Cayman Islands and Providencia. It is a casual visitor to south Florida. The Thick-billed Vireo is generally brownish-green above with a grey green crown and hindneck, the lores are black, the iris dark. There are two white wingbars, and the eyes are surrounded by bright yellow spectacles. The underparts vary from yellow to grey. Curious and tame, it is a species of dense shrubbery and undergrowth. Usually forages in pairs, feeding on insects and berries in low vegetation.

JAMAICAN VIREO *Vireo modestus* 12cm

A vireo which is endemic to Jamaica, where it is, however, a common and widespread bird. The plumage is dull green above and the underparts are pale yellow, and there are two greenish-white wing-bars. The iris is whitish. While foraging, it flicks its tail upwards with regularity as it searches for its diet of insects and fruit. It is a secretive species but also an active one, frequenting brushy edges and roadsides at all elevations.

CUBAN VIREO *Vireo gundlachii* 13cm

A Cuban endemic, this vireo is dark olive-grey above and dull or buffy-yellow on the underparts. There are no wing-bars present, but the side of the face shows a pale yellowish eye-ring. The species is common and widespread throughout the island. Birds are usually found foraging in pairs, but they will also join mixed feeding flocks. It is a sluggish species inhabiting brushland, edges and deep scrub, but also frequents the mountains.

PUERTO RICAN VIREO *Vireo latimeri* 13cm

Endemic to western Puerto Rico. This common vireo has the throat and breast pale grey, and the abdomen pale yellow, which gives it a two-toned appearance. It has an incomplete eye-ring, and there are no wing-bars. The Puerto Rican Vireo is usually associated with limestone hill country, but also frequents forests, scrubland and coffee plantations. It forages in pairs for insects and fruits at all levels in the vegetation, but is more commonly observed near the ground.

BLACK-WHISKERED VIREO *Vireo altiloquus* 15–17cm

The Black-whiskered Vireo is a common resident species occurring widely in the West Indies. Its plumage shows olive-green upperparts, white underparts washed with yellowish-green, a white superciliary stripe, a dark line through the eye and a black malar stripe. There are no wing-bars. It feeds on insects and fruits, which it sometimes takes on the wing. The habitat is forest of all types and at all altitudes.

HISPANIOLAN PALM CROW *Corvus palmarum* 34–43cm

A large black bird with purplish and bluish gloss, endemic to Hispaniola. Although it is smaller than White-necked Crow, its wings are shorter. This causes the palm crow to flap more often. The call is distinctive and crow-like compared to the parrot-like voice of White-necked Crow. Frequents mountain pine forests, and lowlands. Flocks of 30–50 birds are not unusual. Feeds on fruits, seed, insects, snails and lizards. It is relatively tame and allows a close approach, and flicks its tail when calling.

CUBAN CROW *Corvus nasicus* 40–48cm

A crow with a purple gloss. Endemic, common and widespread in Cuba and Caicos and neighbouring islands of the southern Bahamas. The nostrils are not covered by feathers as they are in Cuban Palm Crow *C. minitus*. Its calls are parrot-like, and consist of diverse phrases. On Cuba it frequents open forest, palm plantations, swamp borders, agricultural land and dumps. On Caicos it is found in gardens and agricultural areas and is attracted to fruiting trees. Feeds on fruits, corn, seeds, reptiles, frogs and insects. Forages in small noisy parties, and adapts to deforestation.

CARIBBEAN MARTIN *Progne dominicensis* 17cm

The Caribbean Martin is a breeding species of Mexico, and the West Indies; it only a vagrant in Cuba (the similar Cuban Martin takes its place), the Bahamas, and the Caymans. It winters in northern South America. The male is glossy-blue on the upperparts, head, throat, and flanks; the remainder of the underparts are white. The female has a brownish wash on the belly. Frequents open and semi-open areas near water; seacoasts, urban areas, and cliffs are preferred. The diet consists entirely of insects which are taken on the wing. Will follow cattle to obtain insects which are flushed by them.

TREE SWALLOW *Tachycineta bicolor* 12–15cm

Male

Female

Blue-green above and white below, with a slightly forked tail. A common non-breeder, but present all year, on Cuba and the Caymans, and a migrant elsewhere in the West Indies. Frequents wooded and open areas, especially wetlands. As it rapidly flies and glides in a straight path, it may veer off to capture an insect, but soon returns to its direct flight. Feeds on the wing, and will take berries and seeds when insects are scarce. Large flocks can occur at abundant food sources.

BARN SWALLOW *Hirundo rustica* 15–19cm

A common migrant in the West Indies, but can be observed in all months of the year. Adults are shiny dark blue above, and tan below, with a reddish-brown throat. The tail is deeply forked. A swift flyer, and very acrobatic as it feeds near the ground in open areas. Feeds exclusively on insects which it snatches from the air while in flight. It has a tendency to follow cattle and other mammals which can stir insects into flight. Often skims still water to glean insects from the surface.

ZAPATA WREN *Ferminia cerverai* 16cm

A rare Cuban endemic which is found only in sawgrass marshes, within the area of the Zapata Swamp. It is a brown bird striped with black; the underparts are grey. The legs, bill and tail are long, the wings are short and rounded. A wonderful songster, it will come from a great distance to a playback of its call. This secretive wren usually remains hidden in the dense sawgrass. It usually walks rather than flies, and feeds on insects, spiders, snails, lizards and berries which it takes from the vegetation.

HOUSE WREN *Troglodytes aedon* 11–13cm

The plumage varies from light in St Lucia and St Vincent to very dark elsewhere, and the bill appears longer than in continental birds. A brown bird which has a pale eye-stripe and long tail; upperparts are barred black. In the Lesser Antilles it is common only on Dominica and Grenada where it is a species of undergrowth adjacent to rainforest, and lowland coastal areas. The Lesser Antillean form does not cock its tail. Feeds on insects gleaned from the understorey. Threatened by the introduced mongoose and parasitism by the cowbird.

RUFOUS-THROATED SOLITAIRE *Modistes genibarbus* 19cm

Hector Galvez

The Rufous-throated Solitaire is a multi-island endemic, found in Jamaica, Hispaniola, Dominica, Martinique, St. Lucia and St. Vincent. The upperparts including the head are slate-grey, with darker wings and tail. The outer tail feathers are white, the underparts are mainly grey, the throat, lower belly and undertail coverts are rufous-orange. The beak is small and black, the legs yellow-orange. The St. Vincent form is black above, with olive uppertail. The beautiful flute-like song, like a squeaking gate, is unmistakable. Frequents mountain rain forest, where it forages on insects and berries.

CUBAN SOLITAIRE *Myadestes elisabeth* 19cm

The Cuban Solitaire has a wonderful high-pitched flute-like song which can be heard from a great distance. An endemic to Cuba, and occurring only locally; it is threatened by deforestation. It has olive-brown upperparts and a long tail; the underparts are pale grey. It has a white eye-ring and outer tail-feathers, and a pale malar stripe; the bill is small. It sits motionless on a favourite perch, from where it flies out for insects and fruits like a flycatcher. Frequents dense humid mountain forest.

COCOA THRUSH *Turdus fumigatus* 23cm

In the West Indies this species occurs only on Grenada and St Vincent, where it is common. It is a widespread species of South America, including Trinidad. Plumage of the upperparts is rich brown, and the underparts are paler, with the throat and undertail-coverts being white. The bill is dark. It frequents mountain forests and clearings, as well as cocoa plantations, where it forages on the ground for its food of insects and berries.

BARE-EYED ROBIN *Turdus nudigenis* 23cm

Dark grey above and brownish-grey below. The throat is streaked, and there is a bare yellow eye-ring. In the West Indies it is common on Martinique, St Lucia, St Vincent, the Grenadines and Grenada. Frequents open forest and woodland, secondary growth, plantations and gardens. It is a highly aggressive species and may have been responsible for the decline of the Forest Thrush on St Lucia. Though it will feed on insects and fruits on the ground, it is chiefly an arboreal bird.

WHITE-EYED THRUSH *Turdus jamaicensis* 23cm

The White-eyed Thrush has a rufous-brown head, a grey-white eye and a black bill. The upperparts are dark grey, and the underparts are lighter with a bright white breast-band. This thrush is a bird endemic to Jamaica, where it inhabits the island's wet forests, coffee plantations and other woodlands at higher elevations; it is uncommon in the lowlands. It feeds on fruit and insects, foraging at all levels from the treetops to the ground.

WHITE-CHINNED THRUSH *Turdus aurantius* 24cm

A thrush endemic to Jamaica, and a common species there. It is dark grey above, and paler below; there is a white diagonal bar on the wing. The bill and legs are orange, and the chin is white. The bird cocks its tail upwards as it hops around in fields and along roadways. Feeds on the ground on insects, frogs, lizards, berries, small birds and mice. Frequents forests, road edges, cultivated areas and gardens at mid to high elevations.

RED-LEGGED THRUSH *Turdus plumbeus* 25–28cm

Bahamas race (left); Cuban race (right)

An endemic of the West Indies, ranging through the northern Bahamas, Cuba, Hispaniola, Puerto Rico, Dominica and the Caymans, where it is common and widespread. A variable species with grey upperparts, reddish legs and bill, a red eye-ring and large white tail-tips. The underparts are extremely variable between island races. Hops along roads at dusk and dawn. Forages for insects in leaf litter, and also feeds on fruits and other invertebrates. Frequents forest at all elevations, thick undergrowth, field edges and coffee plantations, and can sometimes be common around human habitation.

FOREST THRUSH *Cichlherminia lherminieri* 25–27cm

This endemic is uncommon on Montserrat, local on Guadeloupe and Dominica, and rare on St Lucia. Large and stocky, with grey-brown upperparts; the breast is marked with large black-and-chestnut spots, the rest of the underparts are white. Bill and legs are yellow, and there is a large bare yellow patch round the eye. Shy, and sings from a hidden perch. Frequents moist mountain forest where it feeds on insects and berries at all levels. May have declined due to habitat loss, and parasitism by cowbirds.

GREY CATBIRD *Dumetella carolinensis* 23cm

A bird which is entirely grey but for its black crown and chestnut undertail-coverts. Its long tail is cocked frequently as it forages near the ground in the dense undergrowth which it prefers. Its cat-like meow is distinctive. In the West Indies it is present commonly and widely as a non-breeder and migrant in the Bahamas, Caymans and Providencia; it is uncommon on Cuba, Jamaica and St Andreas, and is rare elsewhere. Feeds on fruits, berries and insects.

NORTHERN MOCKINGBIRD *Mimus polyglottos* 23–28cm

Grey above and greyish-white below; wings and upward-cocked tail are distinctly marked with white, which is conspicuous in flight. In the West Indies it is a common bird in the Bahamas, the Greater Antilles, Virgin Islands and Cayman Islands. Frequents open country, towns, settlements, mangroves and semi-arid scrub; it is content around human habitation, and is relatively tame. In the Bahamas it appears to be displacing Bahama Mockingbird. Feeds on insects and fruits, and also on household scraps.

TROPICAL MOCKINGBIRD *Mimus gilvus* 23–24cm

Occurs mainly in the Lesser Antilles; a San Andreas form is sometimes considered to be a separate species. The head and back are grey, the wings are dark and the tail blackish with white tips to the outer feathers. The underparts are white or greyish-white. The species appears to be expanding its range as island deforestation is creating new habitat suitable for it. Frequents open settled country. The food is fruit and insects.

BAHAMA MOCKINGBIRD *Mimus gundlachii* 28cm

Larger than Northern Mockingbird, but not aggressive. The upperparts are brownish-grey, finely streaked with dusky on the head and back, and there are two indistinct wing-bars. The underparts are whitish, streaked with dusky, and the tail is tipped with white. Endemic to the Bahamas and locally to Cuba and Jamaica. Inhabits semi-arid scrub – taller and denser vegetation than Northern Mockingbird – where it forages for insects and fruits. It is becoming scarce on Grand Cayman as the population of Northern Mockingbird increases.

SCALY-BREASTED THRASHER *Margarops fuscus* 23cm

The Scaly-breasted Thrasher is endemic to the Lesser Antilles. Thrush-like, it is grey-brown above and white below, and there is a single whitish wing-bar. The underparts are scalloped with grey-brown; the outer tail-feathers are tipped with white, the eye is yellow, and it has a short, black bill, hooked slightly at the tip. It is a rather shy species of the forest canopy where it is relatively common, and frequents lowland dry forest, as well as rainforest. Feeds on insects and fruits.

PEARLY-EYED THRASHER *Margarops fuscatus* 28–30cm

A West Indian endemic with a discontinuous distribution: Puerto Rico, the Virgin Islands, the Lesser Antilles and central and southern Bahamas. The upperparts are brown; underparts white streaked with brown. It has a large yellow bill, a white eye and large white tips to the tail-feathers. An arboreal species which frequents thickets, woodland, forest and mangroves; nesting in cavities. It also inhabits urban areas. Feeds on fruit, insects and small vertebrates; it is an important predator on eggs and nestlings of other species, especially those of Puerto Rican Parrot.

BROWN TREMBLER *Cinclocerthia ruficauda* 23–26cm

Upperparts are dark reddish-olive, the underparts buffy-brown. It has a very long bill, curved at the tip, and that of the female is longer than that of the male. The iris is orange-yellow. Endemic to the Lesser Antilles. A rainforest species, widespread but uncommon. Occupies dry forest in St Lucia. Easily overlooked as it forages in the understorey, it feeds on insects, berries and tree-frogs. It characteristically droops its wings and vibrates them in the manner of a juvenile begging for food, usually while cocking its tail.

GREY TREMBLER *Cinclocerthia gutturalis* 23–26cm

The Grey Trembler is endemic to St Lucia and Martinique, and is similar to the Brown Trembler, which is also resident. The upperparts are dark olive-grey, the underparts greyish-white. The bill is long and downcurved; the iris is white. Trembles as it droops its wings and cocks its tail. Occurs at all elevations in moist forest where it uses its bill to demolish dead leaf clusters and dead vegetation. Feeds on berries, insects and small vertebrates.

PALM CHAT *Dulus dominicus* 20cm

Upperparts

Underparts

The Palm Chat is the national bird of the Dominican Republic, and is in a family of its own, endemic to Hispaniola. An abundant species of open country with trees throughout the island. The upperparts are greenish-brown, and the underparts yellowish-white heavily streaked brown. The eyes are red. The upper mandible of the heavy bone-coloured bill overlaps the lower, perhaps as an adaptation for carrying large sticks to the huge communal nest which may be a few metres across. The feet are very large. It feeds on flowers, fruits and occasional insects.

YELLOW WARBLER *Dendroica petechia* 12cm

Male

Female

One of the most abundant and well known of the wood warblers, its plumage is variable, but is entirely yellow or gold above and below. Males can have chestnut streaking on the breast; some races have varying amounts of chestnut on the head. A common permanent resident in the West Indies, and a non-breeder on some islands, occupying mangroves and coastal scrub. Nests of the the Yellow Warbler are parasitised by the Shiny Cowbird *Molothrus bonariensis*, which is the source of decline in some races. It maintains feeding territories in winter.

CAPE MAY WARBLER *Dendroica tigrina* 12cm

The Cape May Warbler is a common non-breeding resident of the Bahamas, Greater Antilles and the Caymans, but it is rare elsewhere in the West Indies. Identified by its chestnut cheek, white wing-bar, black streaking on the yellowish underparts and yellow to greenish rump; females are whitish. In all plumages there is a trace of yellow on the sides of the neck. Feeds on nectar and fruit during migration and on the wintering grounds, the thin decurved bill being well suited for this purpose. Aggressively defends its food sources.

BLACK-THROATED BLUE WARBLER
Dendroica caerulescens 13cm

A distinctive warbler. The male in breeding plumage is blue above and white below with a black face, throat and flank-streak. The female is green above and yellow below, with a white spot on the wing which assists in identification. Most of the worldwide population of this species winters in the West Indies, and it is common then in the Bahamas, Greater Antilles and Caymans. Forages in low vegetation, and feeds on nectar, fruit and insects. Males are most common in mature forest, whereas females forage in cut-over areas. More numerous at higher elevations.

Male (above); female (below)

YELLOW-THROATED WARBLER *Dendroica dominica* 13cm

A warbler with a bright yellow throat and a striking black-and-white face pattern. The underparts are white, and there is heavy black streaking on the flanks. It has two distinct white wing-bars. Most in the West Indies are winter visitors, the earliest of the wintering wablers to arrive and the first to leave. Common in the Bahamas, Cuba and the Caymans. A resident form breeds on Grand Bahama and Abaco. A canopy species, it commonly creeps along trunks. In its wintering habitat it prefers pines, palms, mangroves, Australian pine and trees laden with epiphytes.

ADELAIDE'S WARBLER *Dendroica adelaidae* 13cm

A West Indian endemic species which is identified by its yellow throat and breast, bluish back, a prominent yellow eye-stripe and two distinct wing-bars. The distribution is discontinuous through Puerto Rico, St Lucia and Barbuda, where the habitats differ: dry scrub in the lowlands in Puerto Rico (where the species is common), forests in St Lucia, and marshy thickets on Barbuda. An arboreal and insectivorous bird, very responsive to a tape of its voice.

OLIVE-CAPPED WARBLER *Dendroica pityophila* 13cm

Endemic to Cuba and the Bahamas (Grand Bahama and Abaco). This species has an olive-green crown, and a yellow breast which is sharply demarcated from its white underparts; there is medium black streaking along the flanks and on the throat, as well as two white wing-bars. Thus similar in appearance to Yellow-throated Warbler, but does not have an obvious face pattern. It is a pine forest species, dwelling in the canopy, where it forages for insects and spiders.

PINE WARBLER *Dendroica pinus* 13–15cm

This species is a large and heavy-billed warbler; it has two whitish wing-bars. The upperparts are olive and unstreaked; the underparts are yellow, fading to whitish on the abdomen, and there is faint variable streaking on the sides of the breast and flanks. The Pine Warbler is a permanent resident of the Bahamas and Dominican Republic, where it prefers the pine forests, feeding on insects, spiders, seeds and berries – though it may be found in other habitats as well.

PRAIRIE WARBLER *Dendroica discolor* 12cm

This tail-bobbing warbler is pale olive-grey above and yellow below. Variable amounts of black streaking are present on the sides; there is a black crescent below the eye, and most birds show a distinctive face pattern. There are two pale yellowish wing-bars. Outside the breeding season, the entire world population of this species is confined to the West Indies, and it is common then in the Bahamas, Greater Antilles, Caymans and Virgin Islands. Prefers scrubby, open areas, and is fond of secondary growth. Feeds on insects and spiders, occasionally nectar.

VITELLINE WARBLER *Dendroica vitellina* 13cm

This warbler is endemic and common on the Caymans and Swan Island. The plumage is olive-green above and totally yellow below with faint side-stripes, and there is a faint face pattern. One of the wing-bars appears indistinct, and the undertail-coverts are yellow. The bird resembles a large washed-out Prairie Warbler, which is a common non-breeder in similar habitat. Frequents scrubland and thickets. Relatively tame and common in gardens, where it actively flies from the ground into the treetops, taking insects from leaves and twigs; also feeds on nectar and fruit.

PALM WARBLER *Dendroica palmarum* 13cm

A tail-wagging species, which is yellow or whitish below, but has yellow undertail-coverts in all plumages; it has a brownish back, white supercilium and olive rump. There are two forms of Palm Warbler which can be separated in the field, one having white underparts, the other yellow. This warbler is a winter visitor in the Bahamas and Greater Antilles, but rare elsewhere. It is usually observed only on the ground or in shrubby, low vegetation. The yellow form seems to prefer more wooded habitats, but the species has no affinity for palms.

PLUMBEOUS WARBLER *Dendroica plumbea* 12cm

Adult   *1st year*

This very tame species is endemic to Dominica and Guadeloupe in the Lesser Antilles. It has greyish upperparts, a white eye-stripe and two wing-bars. It habitually twitches its tail. First-year birds display an identical pattern, but the upperparts are olive, the eye-stripe and underparts yellowish. The Plumbeous Warbler is abundant in the understorey, where it feeds on insects, fruits and berries. Very responsive to a tape playback of its song, or simply to 'pishing': it will actually alight on a person's outstretched hand.

ELFIN WOODS WARBLER *Dendroica angelae* 13cm

The Elfin Woods Warbler was only discovered in 1971. A secretive species, it was overlooked in the zone of the stunted trees (elfin wood) on the highest ridges of the interior. Frequents the dense canopy of humid mountain forests as well. The white underparts with black streaking are similar to those of the Black-and-white Warbler which it closely resembles. Elfin Woods has an incomplete eye-ring, a totally black crown, white stripe above the eye, and white spots on the ears and sides of the neck. It is a hyperactive species, with a ventriloqual, insect-like song, and navigates quickly through the thick foliage like a mouse. Feeds entirely on insects.

WHISTLING WARBLER *Catharopeza bishopi* 15cm

The Whistling Warbler has a loud piercing song, and can be heard long before it is seen. This species is endemic to St Vincent in the Lesser Antilles. The male has a black hood, upperparts and breast band with white underparts and eye-ring. The immature is paler. Uncommon within its habitat of humid mountain, and elfin forests. Feeds on insects by hopping from branch to branch, hanging, or picking insects from crevices. Forages from low to mid level. Shy and secretive, frequently cocks its tail. Deforestation, and the volcanic activity are the main factors impacting on this species.ngates quickly through the thick foliage like a mouse. Feeds entirely on insects.

ARROW-HEADED WARBLER *Dendroica phaetra* 13cm

The Arrow-headed Warbler is endemic to Jamaica. The head and back are white, finely streaked with black, and this pattern is also continuous over the face, breast and remaining underparts. The undertail-coverts are buffy; also streaked with black. The eye is brown surrounded by a white eye-ring, and there are two white wing-bars. Constantly flicks its tail downward. Feeds at all levels of humid forest, where it gleans insects from leaves. Not found in the dry lowlands or on cultivated land. Similar to Black-and-white Warbler, but doesn't flick its tail.

BLACK-AND-WHITE WARBLER *Mniotilta varia* 13–14cm

Generally found throughout the West Indies, where it is a winter visitor. Its habit of creeping up and down tree trunks and branches is unique among wood warblers. To accomplish this it has developed strong legs and a long bill, an adaptation which may allow it to leave the wintering grounds earlier and return there later than other species. Completely black-and-white colouration and striped crown are diagnostic. Frequents a variety of habitats from cloud forest to woodland border, and coffee plantations. Feeds on insects and spiders, many gleaned from bark crevices.

AMERICAN REDSTART *Setophaga ruticilla* 13cm

Generally a fairly common winter visitor to the West Indies. There are also recent nesting records from Cuba. The male has black upperparts, with orange patches at the base of the flight- and tail-feathers; the female is greenish-grey, with the orange replaced by yellow. A very active warbler that constantly displays its colours by fanning tail and flicking wings. In the West Indies it frequents mangroves and lowland forest. Feeds on insects, berries and seeds, usually on the forest floor or up to mid-level, but sometimes snatches insects from the air. Very curious, and responsive to pishing.

Male (above); female (below)

WORM-EATING WARBLER *Helmintheros vermivorus* 14cm

The Worm-eating Warbler is a common non-breeding resident of Cuba and the Bahamas, uncommon in Jamaica, Hispaniola and Puerto Rico, and rare elsewhere. It has greenish-grey upperparts, a brown crown with a black stripe through it, and another black stripe through the eye. The underparts are white shading to buff. Frequents forested areas with heavy leaf litter, dead leaf clusters and heavy vines, usually in the lowlands and foothills. Feeds almost entirely on insects and spiders which it takes on or near the ground. It is solitary and territorial on the wintering grounds.

OVENBIRD *Seiurus aurocapillus* 15cm

The thrush-like Ovenbird is actually a wood warbler, usually recognised by its habit of walking about as it feeds on the forest floor. It calls and bobs its head, cocking its tail frequently. The upperparts are brown, and it has an orange crown bordered by black stripes, and a bright white eye-ring. The underparts are white, heavily spotted black. A common winter visitor in the Bahamas and Greater Antilles, uncommon to rare elsewhere. Feeds on insects, spiders, earthworms and snails taken from the forest floor. Avoids wet lowland forest, but occurs sometimes in mangroves.

LOUISIANA WATERTHRUSH *Seiurus motacilla* 15cm

Separated from Northern Waterthrush by being greyish-brown above and white below with dark brown streaks, and having a white supercilium which expands behind the eye, and a pure white throat. It prefers fast-flowing streams with forested edges, but is rare along the coast and in mangroves. This wintering species is common in the Greater Antilles, uncommon in the Bahamas, and casual elsewhere. Its bobbing motion is slower and more deliberate than Northern's and it is shyer. Feeds on small molluscs, earthworms, crustaceans and even small fish.

119

COMMON YELLOWTHROAT *Geothlypis trichas* 11–14cm

The male's black face-mask rimmed with white is indicative of this species. The female has no mask, and both sexes sport a bright yellow throat and breast, and a whitish abdomen. A common winter visitor in the Bahamas, Greater Antilles and the Caymans, but rare elsewhere. Common in wet brushy fields, marshes and dense vegetation, where it feeds on insects and other invertebrates. Its distinctive song 'witchity-witchity-witch' is sung from a prominent perch. Perches with tail cocked.

Male (above); female (below)

BAHAMA YELLOWTHROAT *Geothlypis rostrata* 15cm

Endemic to the Bahamas. Considerably larger than Common Yellowthroat and has a heavier bill, paler yellow on the entire underparts and a tan or olive cap; there is yellow on the upper border of the face-mask. It is difficult to distinguish between the two species, since their distributions overlap. It is outnumbered by Common Yellowthroat, which is absent during its breeding season. The Bahama Yellowthroat is rather sluggish, and prefers pine woodland understorey and thickets rather than wet areas. Feeds on insects, berries and occasionally small lizards.

Male (above); female (below)

GREEN-TAILED GROUND WARBLER
Microligea palustris 12–14cm

R. A. Behrstock

The head and neck are blue-grey and the rest of the upperparts green; underparts are white; wings short and rounded. The broken white ring around the red eye can be observed upon close inspection. The Green-tailed Ground Warbler is endemic to Hispaniola. It is local in arid areas at higher altitudes, and is usually found on or near the ground, as it creeps along slowly in dense thickets. Usually silent, but sometimes emits a call which sounds like a young bird in distress. Feeds on insects and spiders.

YELLOW-HEADED WARBLER *Teretistris fernandinae* 13cm

This Cuban endemic occupies the forest scrub of the eastern part of the island. It is easily recognised by the bright yellow head which contrasts sharply with a grey back and whitish underparts. This warbler is often to be found in mixed-species foraging groups, moving about near the ground or at mid-levels in the vegetation, where it feeds on a wide variety of insects and fruit. A common species within its range, it is replaced in western Cuba by the similar Oriente Warbler.

WHITE-WINGED WARBLER *Xenoligea montana* 14cm

R. A. Behrstock

This species is uncommon, and is endemic to mountainous areas of Hispaniola. The upperparts are greenish, the head and tail grey and the underparts white. It has a bright white wing-patch, white outer tail-feathers and a pronounced white line above the eye which extends only to the forehead. Forages in pairs, and feeds on insects and seeds; joins mixed flocks in forest undergrowth at higher altitudes. An endangered species, due to habitat destruction, especially in Haiti. This species is sometimes classified as a tanager rather than as a warbler.

BANANAQUIT *Coereba flaveola* 10–12cm

Bahamas race (above);
St Lucian race (top right);
Grenadan race (right)

One of the most abundant species of the West Indies, rare only on Cuba. Its plumage varies from island to island, being entirely black in St Vincent and Grenada. A white supercilium is the diagnostic feature in most plumages, but look also for yellow on the underparts, a white wing-patch and sharp, downcurved bill. Frequents all habitats from lowlands to mountains. It punctures the bases of flowers to obtain nectar, and also feeds on fruit and insects. Acrobatic, sociable and noisy. Common in urban areas.

PUERTO RICAN TANAGER *Nesospingus speculiferus* 16cm

This tanager is a rather nondescript species which is endemic to Puerto Rico. The plumage is olive on the upperparts, and white below with pale brown streaking on the breast. In the field the bird's best feature is the white spot on the wing. It inhabits areas of humid forest and secondary growth above 200m and has a range extending upwards into the mountains. Forms noisy flocks, which feed in the forest canopy. Birds gather into flocks to roost, preferring palm trees or bamboo thickets.

BLACK-CROWNED PALM-TANAGER
Phaenicophilus palmarum
GREY-CROWNED PALM-TANAGER
Phaenicophilus poliocephalus 16–18cm

Black-crowned Palm-tanager. Right: Grey-crowned Palm-tanager

Both species are endemic to Hispaniola. The Black-crowned occurs in the Dominican Republic and eastern Haiti, where it is common in all lowland habitats including urban areas. The Grey-crowned is near endemic to western Haiti, where it is common in areas which have not been deforested, from sea to cloud forest. The Black-crowned has a black head, sharply demarcated from a white throat which blends into the grey underparts, a grey hindneck, and olive-brown upperparts. The Grey-crowned has a black mask, a grey crown and hindneck; there is a sharp contrast between white throat and grey breast. Both species feed on insects, fruits and seeds.

WESTERN STRIPE-HEADED TANAGER *Spindalis zena* 14cm

Male *Female*

Smallest of the stripe-headed tanagers, endemic to the Bahamas, Cuba and the Caymans. The male is black-headed, having a white supercilium and malar stripe; the breast is yellow, the remainder of the underparts white, and there are bright chestnut patches on nape and rump. The female is greyish-olive above and whitish below. Occurs in brushland, coppice, open woodland and secondary growth, and may perch high in the canopy. Feeds on berries, tender leaves and other plant material; insects are apparently not important in the diet. The male is very aggressive during the breeding season.

JAMAICAN STRIPE-HEADED TANAGER
Spindalis nigricephalus 18cm

Male *Female*

A larger and chunkier bird than the other species of stripe-headed tanager. Both sexes have orange-yellow underparts. The male has a black head with a very prominent supercilium and malar stripe, and shows much white on the wings. The female is olive with the breast grey. Endemic to Jamaica where it is widespread. Frequents fruiting trees, and shrubs of the forest and its edges. Feeds on tender leaves, seeds and fruits such as those of royal palm, oranges and pimentos; insects seem not to be significant in the diet.

JAMAICAN EUPHONIA *Euphonia jamaica* 11cm

A short-tailed, stubby-billed, chunky species. The male is generally grey-blue with a yellow abdomen. The female is grey on the head and underparts, while the upperparts and flanks are olive-green. Endemic to Jamaica, where it is common in woodland at all elevations, as well as orchards and gardens. Feeds like other euphonias on mistletoe berries, but also on fruits, seeds, buds and flowers. Moves about the island seasonally; roosts in the canopy in large groups. Sluggish, but acrobatic when feeding. May drive other species from fruiting trees.

Male (above); female (below)

ANTILLEAN EUPHONIA *Euphonia musica* 10–12cm

A West Indian endemic, but variable in appearance over its range from Hispaniola and Puerto Rico through the Lesser Antilles. Chunky, with a short bill and tail. The male is very colourful, having a blue hood and hindneck, and black tail and wings; underparts, rump and forecrown vary from bright yellow to dull orange. The female is olive-green above and yellow below with a blue hood. Forages in dense canopy of mountain forests, feeding mainly on mistletoe berries, other fruits and buds. Forms flocks outside the breeding season.

125

LESSER ANTILLEAN TANAGER *Tangara cucullata* 14–15cm

The Lesser Antillean Tanager, or Hooded Tanager as it was formerly known, is endemic to St Vincent and Grenada in the Lesser Antilles. The male is dull yellow-orange above, the wings and tail are iridescent blue-green, and it has a brown cap. The female is paler with greenish upperparts. The St Vincent form has a smaller bill and a paler crown than the birds on Grenada. An uncommon species in all habitats, foraging in small flocks. It feeds on fruit and insects.

CUBAN BULLFINCH *Melopyrrha nigra* 14–15cm

Male   *Female*

The Cuban Bullfinch is endemic to Cuba and Grand Cayman Island. The male is deep black, and has a thick curved bill, together with a white line on the closed wing edges. The female is paler and ranges from grey-black to olive-grey. A common species, which usually occurs in small flocks. Forages actively at all levels, searching for seeds, buds and fruits. Frequents forests, brushy areas and even mangroves, and is especially fond of dry thickets. A shy and retiring bird.

CUBAN GRASSQUIT *Tiaris canora* 12cm

B. Hallett

Endemic to Cuba, but declining in numbers due to its desirability as a cagebird, and the loss of forest habitat. Male has a bright yellow crescent on the face separating it from the black breast, and the remainder of the upperparts are olive-grey; the female is paler. Appears in flocks with Yellow-faced Grassquit. Feeds on seeds, mainly on the ground, and prefers semi-arid scrub, field borders, and edges of woodlands including pine. The species was introduced to New Providence in the Bahamas in 1963, and is now common there.

YELLOW-FACED GRASSQUIT *Tiaris olivacea* 12cm

This species of grassquit occurs in the West Indies on the Greater Antilles and the Cayman Islands. The male's main plumage feature is a black breast contrasting with a yellow throat and supercilium. It sings from grass spikes. The female is a less striking, mainly yellowish-olive bird, and has the supercilium showing more faintly. Forages on small seeds which are taken while they are still on the stalks. Frequents open grassy areas into the mountains, but is not found in forests. Abundant. Is usually seen in small groups.

BLACK-FACED GRASSQUIT *Tiaris bicolor* 12cm

Male

Female

A common resident over much of the West Indies, but rare and local in Cuba, and absent from the Cayman Islands. The head and underparts of the male are black, while the female is dull olive-brown instead. Occurs in all open habitats, including urban areas of Puerto Rico and Jamaica, and is a relatively tame species. Feeds primarily on the seeds of weeds and grasses, and sometimes on berries. Usually forages in pairs or in small groups.

YELLOW-SHOULDERED GRASSQUIT
Loxipasser anoxanthus 10cm

Male

Female

An endemic to Jamaica. The male has a black head and breast which fades to grey on the abdomen; the undertail-coverts are rusty, and the upper back and wing-coverts bright yellow. The female is brown with rusty undertail-coverts and yellow wing-coverts. A common bird within its range, frequenting foothills and mountains, rarely at lower elevations, usually in dry limestone woodland. Forages in shrubbery, forest edges and gardens, where it feeds on seeds and berries. Usually observed in family groups.

GREATER ANTILLEAN BULLFINCH *Loxigilla violacea* 15–18cm

Endemic to the larger islands of the Bahamas, Jamaica and Hispaniola. Adult males are black with a small red supercilium, throat and undertail-coverts; the female is paler. Immature birds are brown, later fading to grey; the red areas are evident even in these plumages. All have a heavy, cone-shaped bill. Mainly in mountainous areas, uncommon lower down. Feeds on seeds of native and cultivated plants, fruits, green shoots and buds. Declining in areas where there is polluted irrigation water and heavy fertiliser use. Shy, but sometimes sings from an exposed perch. Frequents dense vegetation.

Male (above); female (below)

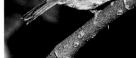

LESSER ANTILLEAN BULLFINCH *Loxigilla noctis* 14–16cm

Male

Female

Endemic to the Lesser Antilles. The male is black with a red chin and throat, but only a small spot of red is evident in front of the eye. The female is brownish-olive above, grey below with orange undertail-coverts; the males in Barbados retain this plumage. Relatively tame, it frequents scrub, understorey, thickets, gardens and occasionally mangroves. Occupies habitats from sea-level to mountain top. Feeds on fallen and other fruit, seeds and insects, and is attracted to scraps.

ORANGEQUIT *Euneornis campestris* 14cm

Female

Male

The Orangequit is endemic to Jamaica, and has a sharp, downcurved bill. Males are mainly blue-black, with a chestnut throat and a black lore. The female has a grey head and back, the remainder of the upperparts are olive, and the underparts are greyish-white. The habitat is open woodland and forest edges, where the birds feed on nectar and fruit, and readily come to feeders; they will also take sap from holes drilled by Yellow-bellied Sapsuckers *Sphyrapicus varius*.

ST LUCIA BLACK FINCH *Melanospiza richardsoni* 13–14cm

Male

Female

The male St Lucia Black Finch is black, with pale pink legs. The female has a grey crown which contrasts with the brown back, and is buffy on the underparts. Both sexes have a heavy cone-shaped bill. Habitually twitches its tail. Can be confused with Lesser Antillean Bullfinch. Uncommon and endemic to St Lucia, it is a ground-dweller, feeding in the leaf litter of humid and dry forests. This species is thought to be the link between the true finches and Darwin's finches of the Galapagos.

ZAPATA SPARROW *Torreornis inexpecta* 16.5cm

The only sparrow endemic to the West Indies, and occurs only on Cuba, where there are three races with restricted distributions scattered across the island. This large sparrow has yellow underparts, a white throat and a dark malar stripe. The crown is chestnut; upperparts olive-grey, and the wings are short and rounded. The Zapata race lives in sawgrass with scattered shrubs, the Guantánamo race lives in arid country and the commonest Cayo Coco form frequents swampy areas and woodland. Feeds on seeds, berries and flowers, and in the rainy season eats insects and snails.

RUFOUS-COLLARED SPARROW *Zonotrichia capensis* 13–16cm

In the West Indies this sparrow is restricted to the mountainous country of Hispaniola, where it is common in both the Dominican Republic and Haiti. A handsome species, the adult has a grey head which is streaked with black; there is a chestnut collar on the hindneck and a black band which separates the throat from the breast. Frequents altitudes above 1,000m in thickets and forest borders. Feeds from the ground on grass and weed seeds, fallen grain and some insects. Common in settled areas.

LESSER ANTILLEAN SALTATOR *Saltator albicollis* 22cm

The Lesser Antillean Saltator was formerly a subspecies of the Streaked Saltator of South America, but now enjoys full species status. It is endemic to Guadeloupe, Martinique, Dominica and St Lucia. A large arboreal finch, with a stout yellow-and-black bill. The upperparts are dark yellowish-green, and the whitish underparts are streaked with olive; there is a black malar stripe and a white supercilium. Frequents edges of forests and pastures, especially in arid areas. A common but shy species, feeding on large seeds and a wide range of fruits.

RED-WINGED BLACKBIRD *Agelaius phoeniceus* 19–23cm

In the West Indies, it is common in the northern Bahamas but absent elsewhere. The male is an unmistakable bird – shiny black, with a large scarlet shoulder-patch – which contrasts sharply with the largely brown female, whose underparts are white and the throat pink in this Bahama race. Breeds in large numbers in freshwater marshes. When feeding – taking seed, fruits, insects and even small invertebrates – birds frequent cultivated land and pastures as well as marshes.

Male (above); female (below)

YELLOW-SHOULDERED BLACKBIRD
Agelaius xanthomus 20–23cm

Endemic to the south-west corner of Puerto Rico and to Mona Island between Puerto Rico and the Dominican Republic. The sexes are alike: glossy black with yellow shoulder-patches. It is an endangered species due to parasitism by cowbirds, and general loss of habitat, and was formerly much more widespread. In Puerto Rico it presently nests only on mangrove islands off the south-west coast, and a small declining population exists near Ceiba. Forages in the canopy and on the ground, taking arthropods and nectar.

EASTERN MEADOWLARK *Sturnella magna* 23cm

A common resident on Cuba, which constitutes its entire distribution in the West Indies. The upperparts are streaked brown, the underparts bright yellow; there is a black V on the breast, and white outer tail-feathers. Frequents open areas, cultivated land, grassland and savanna. Perches on fence posts, wires and other low perches, occurring at all elevations at which suitable habitat exists. It is a ground-feeder, taking mainly insects and other animal material. Has a distinctive wobble as it walks along the ground.

CUBAN BLACKBIRD *Dives atroviolacea* 25–28cm

C. G. Bradshaw

A species endemic to Cuba, where it is abundant and enjoys an extensive range. The Cuban Blackbird's plumage is black with a purple and greenish gloss, the tail is standard size and the sexes are alike. It gathers into large mixed flocks with grackles and other blackbirds, and is considered a crop pest. Frequents farmland, urban areas and forest edges. Feeds on insects, animal feed, seeds and fruits. Has been observed removing external parasites from domestic animals.

GREATER ANTILLEAN GRACKLE *Quiscalus niger* 25–30cm

The male Greater Antillean Grackle has a long tail held in a V-section, and its black plumage reflects bluish. Both sexes have a yellow eye, but the female is smaller and duller than the male. It is endemic and common in the lowlands of the Greater Antilles and Cayman Islands, though the Cayman Brac population is extinct. In Hispaniola it occurs in the highlands. A common species of urban areas, as well as farmland. Feeds principally on insects, but also on food scraps, which are boldly taken from human habitations.

CARIB GRACKLE *Quiscalus lugubris* 24–28cm

A resident species in the Lesser Antilles where it is common and widespread. It is a totally black species with bluish iridescence, and a long V-form tail. The female is smaller than the male, and the tail shorter; both sexes have a yellowish eye. Female plumage varies from island to island. Frequents agricultural land, secondary growth and urban areas. A gregarious, colonial species which feeds on insects, fruit and scraps gleaned from nearby human habitation.

Male (above); female (below)

BLACK-COWLED ORIOLE *Icterus dominicensis* 20–22cm

A resident of Cuba, Hispaniola, Puerto Rico and Andros Island in the Bahamas; presently it is also a rare resident of Abaco, but is severely threatened by unknown factors. The Andros population is threatened by the invasion of Shiny Cowbird *Molothrus bonariensis*. Feeds on insects, fruits, flowers and nectar; frequents lowland forests and secondary growth, anywhere where palms are present. This oriole is black, with large yellow patches on the wing, yellow rump and undertail-coverts. A noisy and active species.

ST LUCIA ORIOLE *Icterus laudabilis* 20–22cm

A species which is an endemic to St Lucia where it is now becoming rare due to the affects of pesticides, parasitism by the Shiny Cowbird *Molothrus bonariensis* and the destruction of habitat suitable for it. The male is mainly black with contrasting orange-yellow areas on the wing, abdomen, undertail-coverts and rump; the female is an overall brown bird. The favoured habitat is rainforest, coastal vegetation and dry scrub, where birds forage for their diet of fruit, nectar and insects.

MONTSERRAT ORIOLE *Icterus oberi* 20–22cm

Female   *Male*

Endemic to Montserrat in the Lesser Antilles. The male is black above, including the wings, with an orange-yellow rump; the black breast is sharply demarcated from the orange-yellow abdomen. The female is yellowish-green above, bright yellow below. This oriole was thought to be seriously threatened after the recent eruption of Mt Soufriere rendered over half the island uninhabitable, but a census showed that a population of over 4,000 individuals remained. The diet is almost exclusively insects, gleaned from the underside of leaves. Frequents mountain forests, near stands of heliconias.

MARTINIQUE ORIOLE *Icterus bonana* 18–21cm

Phillippe Feldmann

This oriole is a rare but widespread endemic of Martinique. It is considered endangered, and has declined most in the northern part of the island, in part due to parasitism by the Shiny Cowbird *Molothrus bonariensis*. The head, neck and upper breast are orange-chestnut, the rest of the underparts and the rump are orange-red; wings, mantle and tail are black. The habitats occupied are as diverse as rainforest, semi-arid hills and gardens, but the Martinique Oriole is essentially a bird of the canopy, feeding on insects, and supplementing them with fruit.

JAMAICAN ORIOLE *Icterus leucopteryx* 21cm

Endemic to, and a common resident of, Jamaica and St Andrew's Island; it was formerly also present on the Caymans but is now extinct there. An unmistakable pale yellowish-green oriole with a black face and bib, and large white wing-patches; the St Andrew's race is a brighter yellow. Found commonly in all habitats except mangroves, from the coast to the mountains; a common garden bird. Pries away loose bark as it forages for insects; also feeds on fruits and flowers.

WHITE-WINGED CROSSBILL *Loxia leucoptera* 15cm

Female   *Male*

This form of the White-winged Crossbill occurs only in the mountains and pine forests of Hispaniola; the species otherwise has a much more northerly distribution, and this West Indian population should be a candidate for being considered a separate species – as has been done recently with various populations of Red Crossbill. The male is pale red overall with two white wing-bars, while the female is dusky and yellow-rumped. Both sexes have crossed mandibles used in extracting pine seeds from the cones.

ANTILLEAN SISKIN *Carduelis dominicensis* 11cm

Endemic to Hispaniola, and as such is a common denizen of mountain pine forests of the Dominican Republic – but rare in Haiti. The male has a distinctive black head, yellowish body and black tail; the female is olive-green above and drab below with two pale wing-bars. Forages in flocks in the forest, travelling from tree to tree, but it is also a species of forest edge and grass fields, and sometimes frequents cultivated areas. Active and acrobatic, feeding on various seeds and possibly on insects.

GLOSSARY

arboreal A species which lives in and about trees
brackish Slightly salty
canopy Treetops
coniferous Pertaining to evergreens, especially pine
coppice A thicket of small trees and bushes
crepuscular By twilight
crown The treetop
elfin forest A stunted mountain forest
epiphyte A plant which grows on another plant
endemic A species confined to a specific area, being found nowhere else in the world
emergent vegetation Plants whose roots are submerged, and the foliage emerges above the water
flight-feathers The long feathers of the wings
frugivorous Feeding exclusively on fruit
glean Gather
insectivorous Feeding exclusively on insects
lore The area between bill and eye
mask A distinctively coloured patch of feathers around the eye
migrant A species which takes long and regular journeys between its breeding and non-breeding areas
moustache or malar stripe Lines which extend down and back across the face from the base of the bill
nape That point where the back of the head merges with the neck
omnivorous Feeds on a variety of animal and plant material
pelagic Oceanic
plunge-diving Diving for food from a height, submerging completely in the process
primaries The flight-feathers of the outer wing joint
resident Present in a general area for the whole of the year
sallies Leaping or flying from a perch to feed
scapulars Feathers which cover the shoulder
secondaries Flight-feathers of the inner portion of the wing
supercilium A stripe over the eye
understorey The forest area below the canopy
undertail-coverts Small feathers which cover the bases of the tail feathers beneath the tail
uppertail-coverts Small feathers which cover the bases of the tail feathers above the tail
wing-coverts Small feathers which cover the bases of the primaries and secondaries

FURTHER READING

Bond, J. *Birds of the West Indies*. HarperCollins, London, 5th Edition 1985

Bradley, P. *Birds of the Cayman Islands*. Caerulea Press, Italy, 1995

Brundell-Bruce, P. *The Birds of New Providence and the Bahama Islands*. Collins, London, 1975

Dod, A. *Endangered and Endemic Birds of the Dominican Republic*. Cypress House, Ft. Bragg CA, 1992

Downer, A and Sutton, R. *Birds of Jamaica*. Cambridge University Press, Cambridge, 1990

Evans, P. *Birds of the Eastern Caribbean*. Macmillan Education Ltd., London, 1990

Garrido, O. and Kirkconnell, A. *Fieldguide to the Birds of Cuba*. Cornell U. Press, Ithaca NY, 2000

National Geographic Society, *Field Guide to the Birds of North America*, 3rd Edition. National Geographic Society, Washington DC, 1999

Raffaele, H. *A Guide to the Birds of Puerto Rico and the Virgin Islands*. Princeton U. Press, NJ, 1989

Raffaele, H. et al. *A Guide to the Birds of the West Indies*. Christopher Helm, London, 1998

Wauer, R. *A Birder's West Indies*. University of Texas Press, Austin, 1992

White, A. *A Birder's Guide to the Bahama Islands Including the Turks and Caicos*. ABA, Colorado Springs CO, 1998

Extralimital Islands

Amos, J. *A Guide to the Birds of Bermuda*. Corncrake-Warrick, Bermuda, 1991

ffrench, R. *A Guide to the Birds of Trinidad and Tobago*. Christopher Helm, London, 2nd Edition 1992

Voos, K.H. *Birds of the Netherland Antilles*. DeWalberg Press, Curacao, 1983 (English version)

INDEX

141

Acknowledgements

The photographer wishes to thank a myriad individuals, departments and staff for their assistance in my pursuit to locate the unique avifauna of the West Indies.

The Bahamas: Eric Carey of the Dept. of Agriculture. Cuba: Al and Chino Garcia, and Arturo Kirkconnell. Cayman Islands: Patricia Bradley and Nancy Norman. Jamaica: Ann Hayes-Sutton and Herlitz Davis and the rest of the gang at Windsor Caves. Haiti: Glenn Smucker for being in the right place at the right time. Dominican Republic: Kate Wallace and all the members of Club de Observadores de Aves Annabelle Dod. Puerto Rico: Sharon Goldade of the U.S. Fish and Wildlife Service, Wayne Arendt, Miguel Canals and Adrian Muniz at the Departamento de Recursos Naturales y Ambientales. Montserrat: Phil Atkinson, Gerard Grey, 'Scriber' and the remaining staff at Forestry of Environment Division Ministry of Agriculture. Philippe Feldmann for his images from Guadeloupe and Martinique. Dominica: Steven Durand, Bertrand Baptiste and staff at the Ministry of Agriculture and Environment, Forestry & Wildlife. St. Lucia: Donald Anthony and staff of the Forest and Lands Dept., Ministry of Agriculture, Fisheries, Forestry and Environment. St. Vincent: Members of Avian Eyes, especially Amos Glasgow and Fitzroy Springer and the staff at Forestry Division, Ministry of Agriculture and Labor.

Photographs in this book were taken by Allan Sander except where other photographers' names are indicated next to the photographs.